IntechOpen

Supporting open minds since 2005

Stability and Applications of Coordination Compounds
http://dx.doi.org/10.5772/intechopen.83186
Edited by Abhay Nanda Srivastva

Contributors
René Gutiérrez, Oscar Portillo, Gloria E. Moreno, Daniela Gutiérrez, Guadalupe Hernández, Eduardo Brambila, Muhammad I. Iqubal Khan, Saima Gul, Murad Ali Khan, Ashish K. K. Asatkar, Deepali Asatkar, Mamta Tripathi, Vikas Vilas Borge, Raju M. Patil, Jagvir Singh, Abhay Nanda Nanda Srivastva, Netrapal Singh, Anuradha Singh, Xinwen Guo, Yi Zuo, Min Liu, Senthilkumar Muthaiah, Anita Bhatia, Muthukumar Kannan, Yenny Patricia Avila Torres, Saikat Sarkar, Palash Mandal, Uttam Das, Kamalendu Dey

First published in London, United Kingdom, 2020 by IntechOpen
IntechOpen is the global imprint of INTECHOPEN LIMITED, registered in England and Wales, registration number: 11086078, 7th floor, 10 Lower Thames Street, London, EC3R 6AF, United Kingdom
Printed in Croatia

British Library Cataloguing-in-Publication Data
A catalogue record for this book is available from the British Library

Additional hard and PDF copies can be obtained from orders@intechopen.com

Stability and Applications of Coordination Compounds
Edited by Abhay Nanda Srivastva
p. cm.
Print ISBN 978-1-83880-057-4
Online ISBN 978-1-83880-058-1
eBook (PDF) ISBN 978-1-83880-727-6

For EU product safety concerns:
IN TECH d.o.o., Prolaz Marije Krucifikse Kozulić 3, 51000 Rijeka, Croatia, info@intechopen.com or visit our website at intechopen.com.

Meet the editor

Dr. Abhay Nanda Srivastva currently holds a position of Head of the Department of Chemistry, Nitishwar College (B.R.A. Bihar University), Muzaffarpur. He completed his M.Sc. and Ph.D. degree in Chemistry from C.C.S. University, Meerut, India on the topic *"Physico-Chemical Studies on Binuclear Complexes of Transition Metal ions derived from Pyrimidine derivatives and Ketones and their Biological Screening."* Dr. Srivastva has approximately seven years of teaching and research experience in chemistry. Before joining B.R.A. Bihar University, he served as Assistant Professor of Chemistry at the IMS Engineering College, Ghaziabad and IEC College of Engineering & Technology, Greater Noida (Affiliated unit of Dr. A.P.J. Abdul Kalam Technical University, Lucknow, India). Dr. Srivastva is continuously engaged in research-cum-academic activities with publications in journals of international repute and attending or presenting research papers, invited talks at international/national seminar, conferences, symposiums, and workshops.

Contents

Preface

Among various research studies done in chemical sciences by researchers around the globe, coordination chemistry has had a prominent position due to the versatile use of coordination compounds. Interest of researchers in this field is also due to ease of synthesis process of mono-metallic to multi-metallic compounds with flexible, simple lone pair/s donating organic, and inorganic moieties. Engagement of chemists and material scientists in designing potent lone pair rich ligands and their complexes as well in recent years is a ray of hope to enrich our scientific community with high grade metal-based molecules for their utilization in different fields of social wellness. The present compilation is the research findings, thoughts, and scientific views of researchers on coordination compounds with special attention on their stability and wide applications. The efforts of authors are presented here as an edited book and divided in nine chapters, accordingly. The chemical aspects of discussions have been accentuated to a great extent as well as relevant physical and spectroscopic data have also been incorporated. Moreover, simple analogies and graphics have been used to explain concepts in easier way. It is a structured attempt to present the matter in a comprehensive manner so that the personnel from research, academic, and industrial backgrounds are able to enhance their knowledge and expertise about the current situation and future possibilities in this emerging field of chemical research. However, post-graduate level chemistry students may also benefit from this book as the study of coordination compounds has been introduced in the syllabus of nearly all universities in the world. As an editor, I believe that every person dealing with coordination chemistry will be advantaged with this book. Furthermore, I am very obliged to the authorities of Nitishwar Mahavidyalaya (B.R.A. Bihar University), Muzaffarpur, India for providing creative and encouraging environments throughout the progress of this book project. Faculty members of the Post Graduate Department of Chemistry, B.R.A. Bihar University, Muzaffarpur, India also deserve a thank you for fruitful discussion about this project. For the most, I am grateful to Dr. Netrapal Singh, Associate Professor, Department of Chemistry, D.D.U. Gorakhpur University, Gorakhpur, India whose assistance throughout made this job easier. I am also thankful to my family members for their support and patience. Finally, I owe my hearty thanks to all the authors who contributed chapters for the book and also to the entire team of IntechOpen, London, United Kingdom for taking responsibility of publishing this book.

Abhay Nanda Srivastva
Department of Chemistry,
Nitishwar Mahavidyalaya (B.R.A. Bihar University),
Muzaffarpur, India

Dedication

Dedicated affectionately to my lovely children and family: Shivansh, Rudransh and Divyansh.

Chapter 1

Crystal Structure and Solid-State Properties of Metal Complexes of the Schiff Base Ligands Derived from Diacetylmonoxime: A Brief Review

Palash Mandal, Uttam Das, Kamalendu Dey and Saikat Sarkar

Abstract

The fabulous advancement of a large section of modern coordination chemistry depends upon different kinds of strategically designed and functionally tuned ligand systems; Schiff base ligands play a pivotal role among them. Such Schiff bases become more motivating when they are designed to be synthesized using very simple organic molecules. This paper reviews our work on a family of three functionally different types of Schiff base ligands, derived from diacetylmonoxime, which have been employed to synthesize mononuclear metal complexes with various binding modes of ligands and topologies around the metal centers. Such Schiff base ligands have been synthesized by reacting diacetylmonoxime with diethylenetriamine, 1,3-diaminopropane-2-ol, and morpholine N-thiohydrazide. The synthesized Schiff bases and the metal complexes of such "privileged ligands" show many interesting supramolecular coordination architectures involving different weak forces, e.g., H-bonding, C–H···π interactions, etc.

Keywords: Schiff base, diacetylmonoxime, crystal structure, weak force interactions, semiconducting behavior, optical properties

1. Introduction

The synthesis and characterization of metal complexes of Schiff bases have been started since 1865. But the importance of Schiff base ligands in several fields compelled us to consider it as a "privileged ligand" even in recent days [1, 2]. Currently, there has been considerable interest in the chemistry of Schiff base metal complexes, primarily because of their tremendous biochemical activity, viz., antibacterial [3, 4], antimalarial [5, 6], antiviral [7, 8], and antitumor activities [9–11]. Besides, some transition metal Schiff base complexes were found to be efficient catalysts in organic synthesis [12–18]. Such types of Schiff base metal complexes are also very interesting for opening of a new pathway in crystal engineering [19, 20]. Various kinds of supramolecules of diverse fascinating structures are being

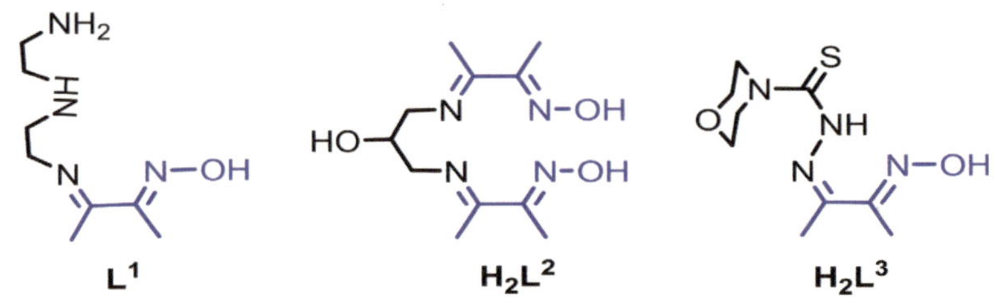

Figure 1.
Structural drawing of the ligand systems.

synthesized by different Schiff bases [20, 21]. Different types of weak force interactions (e.g., H-bonding, π···π, C–H···π, etc.) are responsible for construction of such new metal organic frameworks (MOF) [22–24]. Multinuclear metal complexes of Schiff base are of real importance on the field of magnetochemistry [25–28]. Various ferro- and antiferro-type magnetic interactions are responsible for the generation of different magnetic materials. Besides, the solid-state properties, e.g., variable temperature conductivity, and optical properties of such complexes are also producing very interesting results which are extremely important in materials chemistry. Very recently the comparison of different observed physical properties with the theoretically predicted values is being done by the use of DFT approach [29–32].

In the above context, the development of new pathways for the synthesis of new Schiff base ligands and their metal complexes is of immense significance. The strategic pathway becomes more important when the Schiff base ligands and their corresponding metal complexes are produced in a controlled approach fulfilling the main objectives of the synthesis. We have chosen easily available, exceptionally economical, and full of exciting properties organic molecule, diacetylmonoxime, as our precursor molecule for the synthesis of many new Schiff base ligands by reacting it with different molecular amine systems. One of the main advantages of such Schiff base ligands is the change in their ligational behavior depending on the metallic systems and the stoichiometry.

In this review, we have selected only three Schiff base ligands (**Figure 1**) derived from diacetylmonoxime and three different amine systems. Though a huge number of metal complexes have been synthesized and characterized using such ligand systems, only the structure of the ligands and metal complexes for which single crystal or PXRD have been determined, are discussed in this mini-review including the weak force interactions depicted therein. Some of the solid-state properties, viz., electrical and optical properties of such complexes, are also discussed to enlighten their fascinating material properties.

2. Synthesis of the ligands

2.1 Mono-imine Schiff base (L^1) ligand

The stoichiometrically controlled condensation reaction of diacetylmonoxime (dam) (1.01 g, 10 mmol) and diethylenetriamine (dien) (1.04 g, 10 mmol) in 1:1 molar ratio in methanol (15 ml) on constant stirring for 45 min at room temperature and then refluxing for 2 h on water bath (**Figure 2**) afforded the monocondensed amine-imine-oxime Schiff base ligand 3-((2-((2-aminoethyl)-amino)ethyl)imino) butan-2-one oxime (**L**^{1}). Yield is 55% [29].

2.2 Di-imine Schiff base (H_2L^2) ligand

The condensation reaction of 1,3-diaminopropane-2-ol (0.45 g, 5 mmol) (dapol) with diacetylmonoxime (1.01 g, 10 mmol) in 1:2 molar ratio in methanol (25 ml) under gentle reflux for 2 h yielded the tetradentate bicondensed di-imine Schiff base 3,3′-((2-hydroxypropane-1,3-diyl)bis(azanylylidene))bis(butan-2-one) dioxime (**H_2L^2**) (**Figure 3**). Yield is 65% [33].

2.3 Thio-hydrazone Schiff base (H_2L^3) ligand

The condensation reaction of diacetylmonoxime (1.01 g, 10 mmol) with morpholine N-thiohydrazide (mth) (1.6 g, 10 mmol) in 1:1 molar ratio in ethanol (30 ml) on refluxing for 2 h afforded a gummy mass with very low yield. The isolation of solid ligand in pure form with high yield is still a challenge. Considering the low yield of the ligand, all the complexation reactions with this ligand were carried out under in situ condition. However, the molecular thiol form of the ligand **H_2L^3** (**Figure 4**) was established from the structural analysis (single-crystal X-ray diffraction) of its metal complexes [34–36].

A slight modification of synthetic procedure by continuing the refluxing procedure for 16 h with few drops of water in 1:1 molar ratio in ethanol (50 ml) (shown in the left part of **Figure 4**) yielded a light yellow crystalline solid compound (yield 45%). X-ray diffraction study of single crystal along with other analytical data of this compound inferred a zwitterionic structure of a nitrogen–sulfur heterocyclic compound, N-(3,4-dimethyl-1,2,5-thiadiazole-2-ium-2-yl)morphine-4-carbathio-ate (abbreviated as **L^4**) [25]. Metal complexes with **L^4** have not been reported yet.

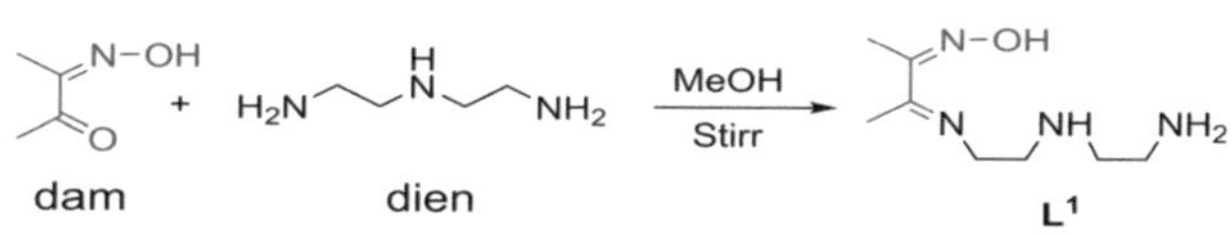

Figure 2.
Scheme of formation of ligand L^1.

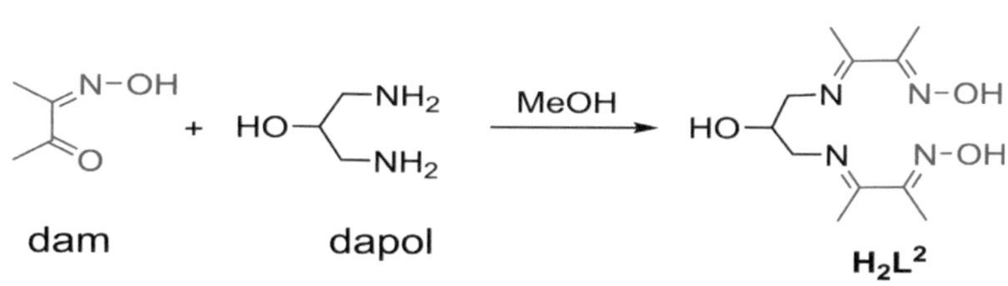

Figure 3.
Scheme of formation of ligand H_2L^2.

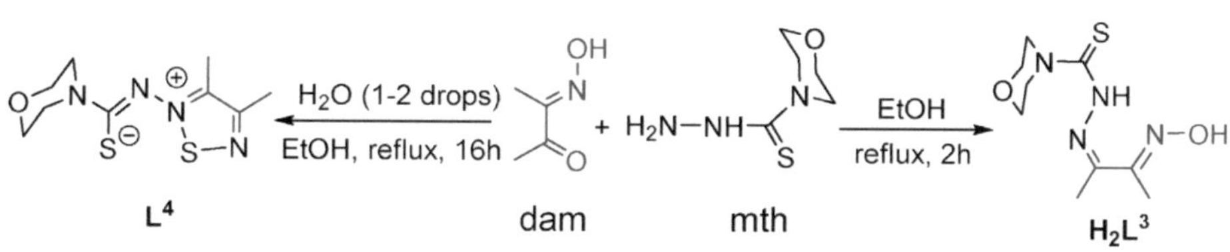

Figure 4.
Scheme of formation of ligand H_2L^3 and zwitterionic heterocycle (L^4).

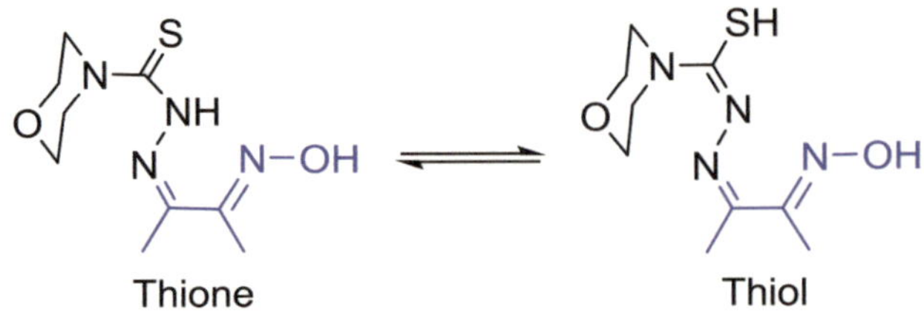

Figure 5.
Thione-thiol tautomerism.

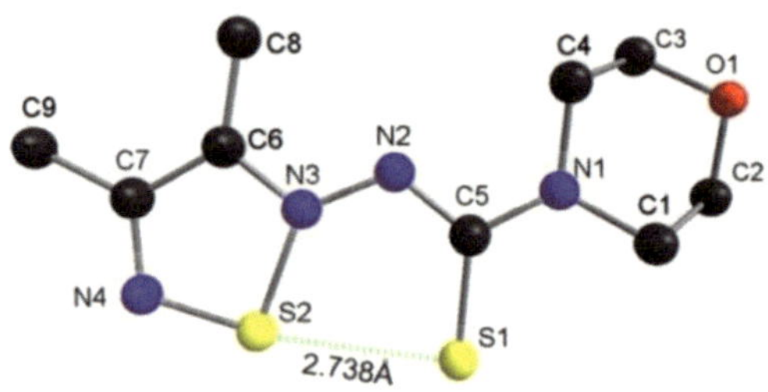

Figure 6.
Perspective view of the ligand $\boldsymbol{L^4}$ *with atom number scheme (hydrogen atoms are omitted for clarity).*

The hydrogen atom attached with N atom of hydrazide group can undergo thione-thiol tautomerism (**Figure 5**). Thus NNS coordination mode is facilitated during the formation of complexes [34].

2.3.1 Crystal structure of $\boldsymbol{L^4}$

The molecular structure of $\mathbf{L^4}$ was established with the single-crystal X-ray diffraction studies. It was observed that the N–N, N–C, and C–S bond distances are shorter than the corresponding single bond distances indicating a partial double bond character due to resonance stabilization using π-bonded electrons. The morpholine moiety preferred chair conformation in the solid state. The distance between the nonbonded $S{\cdots}S^-$ is 2.738 Å, and weak interactions (**Figure 6**) were observed between them in the solid state. A two dimensional supramolecular network is formed by the H-bonding and π- π interactions [34].

3. Synthesis of metal complexes

3.1 Metal complexes with mono-imine Schiff base ($\mathbf{L^1}$) ligand

With this new neutral N_4 donor ligand system, a crystalline Ni(II) complex has been synthesized. Two routes of synthesis of the nickel complex (**complex 1**) were reported (**Figure** 7). According to **Route 1**, the nickel complex was synthesized by in situ reaction of an equimolar mixture of diacetylmonoxime (1.01 g, 10 mmol) and diethylenetriamine (1.04 g, 10 mmol) and $Ni(SCN)_2.4H_2O$ (2.47 g, 10 mmol) in methanol under reflux condition, which produces $[Ni(L^1)(SCN)_2]$ (**complex 1**). In **Route 2, complex 1** was synthesized by following the same steps, but nickel chloride ($NiCl_2.6H_2O$) (2.37 g, 10 mmol) was used instead of $Ni(SCN)_2$ and then by treating the product with NH_4SCN (1.52 g, 20 mmol) in aqueous-methanol (1:1, 10 ml). Pure crystals (single) were obtained from the slow evaporation of the mother liquor. Yields of the **complex 1** in the routes 1 and 2 are 85% and 82% respectively [29].

3.2 Metal complex with di-imine Schiff base (H_2L^2) ligand

The di-imine Schiff base ligand (H_2L^2) was employed for the synthesis of a oxovanadium complex, the PXRD of which have also been determined. Reflux of equimolecular mixture of $\mathbf{H_2L^2}$ (1.28 g, 5 mmol) and vanadyl acetate (0.93 g, 5 mmol) in methanol (30 ml) (**Figure 8**) afforded the greenish-gray complex of vanadium (**complex 2**), having composition [$VO(L^2)$]. In this complex the ligand is found to act as a dibasic N_4 donor system [33].

3.3 Metal complexes with thio-hydrazone Schiff base (H_2L^3) ligand

The thio-hydrazone Schiff base ligand (H_2L^3) is very interesting, and the thiol form of the Schiff base is always observed for the binding purposes (see **Figure 5**) with the metal systems during complexation. Most interestingly, this ligand is found to show two types of binding modes (**Figure 9**), one is observed through N,S donor atoms and another one through N,N,S donor atoms. An efficient control over the ligand for binding through either N,S or N,N,S mode has been achieved though specific choice of metal systems [34–38].

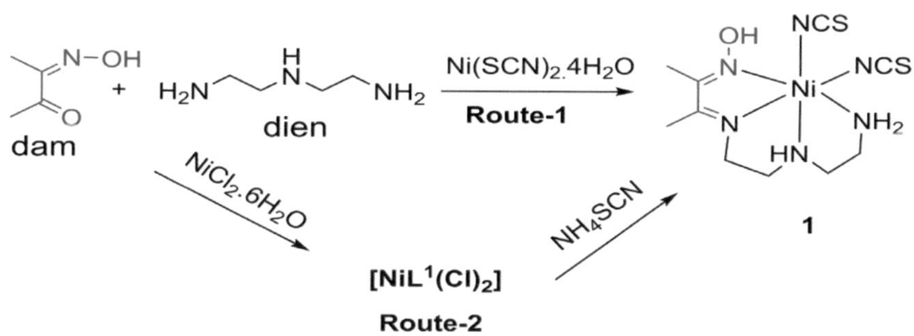

Figure 7.
Scheme of preparation of Ni(II) complexes.

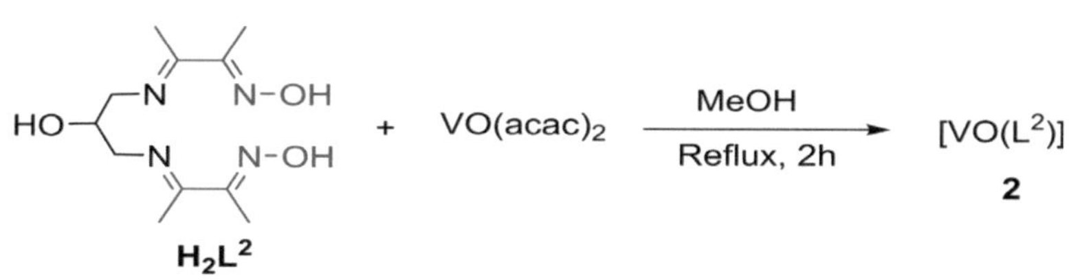

Figure 8.
Scheme of preparation of VO(IV) complex.

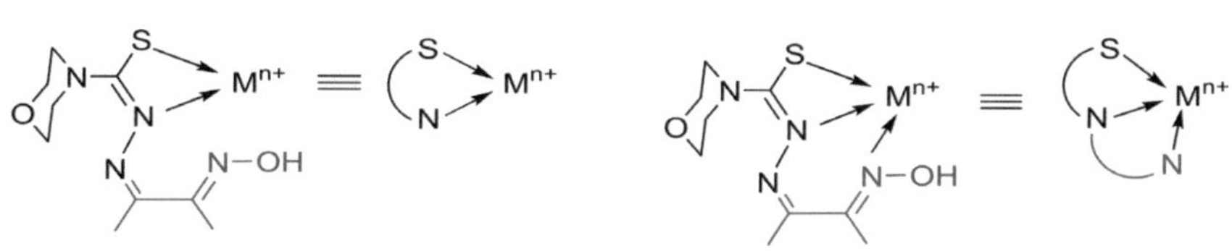

Figure 9.
Different modes of binding of $\mathbf{H_2L^3}$ ligand.

An organometallic complex, [PhHg(HL^3)] (**complex 3**), and the zinc(II) complex [Zn(HL^3)(OAc)(H_2O)] (**complex 4**) were prepared by the gentle reflux of the equimolar quantities of diacetylmonoxime (0.5 g, 5 mmol) and morpholine N-thiohydrazide (0.8 g, 5 mmol) (**Figure 10**) in the presence of the respective metal salts, [PhHg(OAc)] (1.68 g, 5 mmol) and [Zn$(OAc)_2$].2H_2O (1.1 g, 5 mmol). Yields of the complexes are 65% (**3**) and 75% (**4**). The pure single crystals of the complexes were obtained from their chloroform solutions [34, 35].

The Cd(II), Cr(III), and Fe(III) complexes, i.e., [Cd$(HL^3)_2$)] (**complex 5**), [Cr$(HL^3)_2$]Cl.3H_2O (**complex 6**), and [Fe$(HL^3)_2$]Cl.3H_2O (**complex 7**), were prepared by the gentle reflux of the mixture of diacetylmonoxime (0.5 g, 5 mmol) and morpholine N-thiohydrazide (0.8 g, 5 mmol) and the respective metal salts ([Cd$(OAc)_2$].2H_2O (0.67 g, 2.5 mmol), $CrCl_3$.6H_2O (0.67 g, 2.5 mmol), and $FeCl_3$.6H_2O (0.68 g, 2.5 mmol) in the ratio of 2:2:1 (**Figure 10**). Yields of the complexes had been recorded as 75% for each complex. The pure single crystals of the complexes were obtained from their chloroform solutions [35, 36].

Another zinc(II) complex (**complex 8**) was synthesized by the reflux of the mixture of diacetylmonoxime (0.5 g, 5 mmol) and morpholine N-thiohydrazide (0.8 g, 5 mmol) in the presence of Zn$(OAc)_2$.2H_2O (0.55 g, 2.5 mmol) in the molar ratio of 2:2:1 in water–methanol (1:1, v/v) mixture (**Figure 11**). The yellow colored complex of composition [Zn$(HL_3)_2$]. 2H_2O had been separated with 60% yield and recrystallized from chloroform [37].

The complexes of Ni(II) (**complex 9**), Co(II) (**complex 10**), and Cu(II) (**complex 11**) were synthesized by gentle reflux of three sets of equimolar quantities of diacetylmonoxime (0.5 g, 5 mmol) and morpholine N-thiohydrazide (0.8 g, 5 mmol) in the presence of the metal salts (Ni$(OAc)_2$.4H_2O (1.24 g, 5 mmol), Co $(OAc)_2$.4H_2O (1.25 g, 5 mmol), and Cu$(OAc)_2$.4H_2O (1.28 g, 5 mmol), respectively) in ethanol (50 ml) (**Figure 11**) [29]. General molecular formula of the metal complexes **9–11** is [M^{II}(HL^3)(OAc)] where M stands for Ni, Co, and Cu.

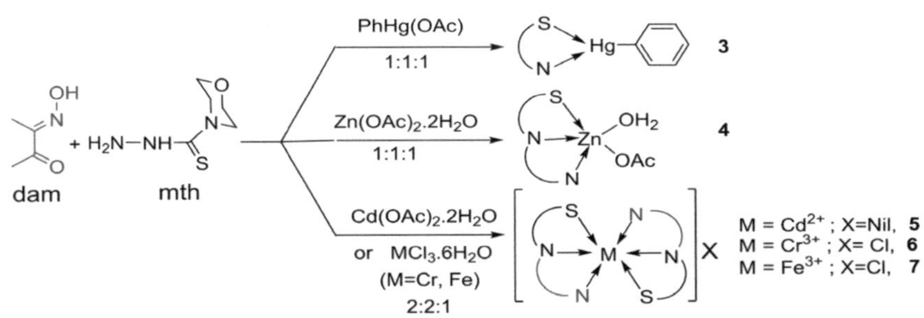

Figure 10.
Scheme of formation of metal complexes using H_2L^3 *ligand indicating different binding pattern.*

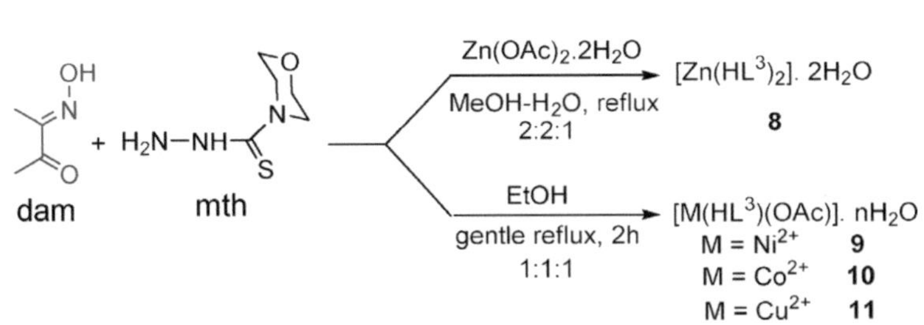

Figure 11.
Scheme of preparation of other metal complexes using H_2L^3 *ligand.*

DOI: http://dx.doi.org/10.5772/intechopen.90171

4. Crystal structures, PXRD, and some interesting properties of the metal complexes

4.1 Crystal structure and catalytic properties of nickel(II) complex with the mono-imine Schiff base ligand (L^1)

The neutral monomeric complex $[Ni(L^1)(NCS)_2]$ (**1**) has been found to possess octahedral geometry where central Ni(II) is coordinated by the neutral ligand $\mathbf{L^1}$ with tetradentate N_4 binding mode and two N-bonded thiocyanate ions occupying the *cis*-position. **Figure 12** depicts the molecular structure of the nickel (II) complex **1** [29].

The coordination environment around the nickel(II) ion is surrounded by N_6 fashion (four N from ligand and two N from thiocyanate ions) tending towards distorted octahedral geometry. The Ni^{2+} center is not lying exactly within the equatorial plane of N_4 moiety, and unequal axial and equatorial bond distances (2.112 Å and 2.072 Å, respectively) confirm the distortion. The non-coordinated O–H groups on the ligand L^1 are engaged in H-bonding interactions with thiocyanate S atoms (**Figure 6**) which lead to 1D supramolecular sheet-like arrangement (**Figure 13**). These H-bonding interactions lead to O···S separations of 3.132 Å and play prominent role in crystal packing.

*4.1.1 Catalytic activity of complex **1***

Analytical grade reagents and freshly distilled solvents, viz., water, acetonitrile, methanol, and dichloromethane, were used to check the catalytic activity. The oxidation reaction was carried out in liquid phase under vigorous stirring in two-necked round bottom flask fitted with a water condenser and placed in an oil bath at 60°C. Substrate (5 mmol) was taken in 10 ml solvent(s) for different sets of reactions along with 5 mg catalyst, to which 10 mmol of *tert*-Butyl hydrogen peroxide

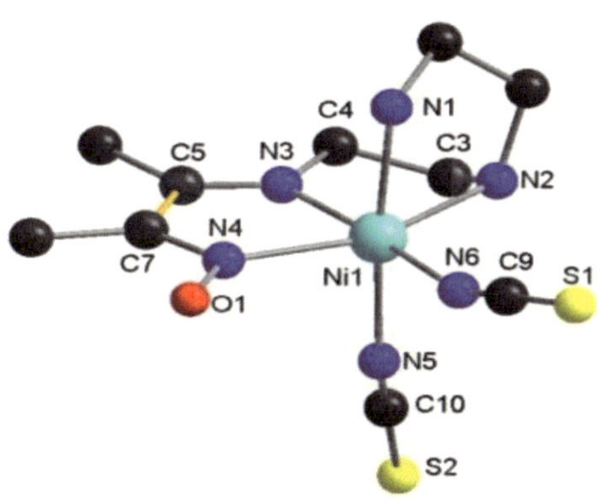

Figure 12.
Perspective view of complex 1 with atom numbering scheme (hydrogen atoms are omitted for clarity).

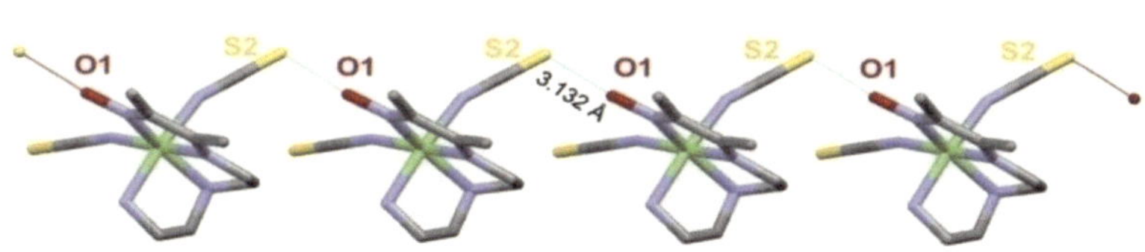

Figure 13.
S···O interaction in the crystal of complex 1 (hydrogen atoms are omitted for clarity).

(TBHP, 70% aq.) was added as oxidant immediately before the start of the reaction. Aliquots of the reaction mixture were withdrawn at various time intervals, and the products were analyzed by using Agilent 6890A gas chromatograph equipped with HP-1 capillary column and FID. All reaction products were identified and estimated by using an Agilent GC–MS (QP-5050 model).

The $[Ni(L^1)(NCS)_2]$ (**1**) exhibits good catalytic activity towards the oxidation of styrene in the presence of TBHP which leads to the formation of benzaldehyde as a major product with small amount of styrene oxide as shown in **Figure 14**. By varying solvents, reaction time, temperature, and substrate-to-oxidant ratio, a successful conversion of 70% of the styrene was reached at the optimum condition [29].

4.2 PXRD structure and solid-state properties of oxovanadium(IV) complex with the di-imine Schiff base ligand (H_2L^2)

Despite our repeated attempts and best effort, the single crystal of the oxovanadium(IV) complex with the ligand H_2L^2 could not be grown, and it led us to carry out the powder X-ray diffraction (PXRD) study to characterize the oxovanadium(IV) complex **2**. The composition of the complex is $[VO(L^2)]$, where the ligand H_2L^2 acts as a dibasic $[(L^2)^{2-}]$ tetradentate arrangement. The result of the PXRD shows that the unit cell is monoclinic in nature and the cell parameters of the complex **2** are a = 16.0619 Å, b = 11.5517 Å, c = 12.7262 Å, and β = 122.427°; and the cell volume (V) is 1993.05 $Å^3$.

Other solid-state properties, viz., electrical, optical, and thermal properties of the complex $[VO(L^2)]$, have also been studied [24]. The complex is electrically an insulator at room temperature; however, the conductivity is increased as the temperature increases from 330 K, indicating the semiconducting nature of the complex. It behaves as an n-type semiconductor, and the semiconducting behavior of the oxovanadium(IV) complex with the dibasic Schiff base ligand was substantiated by the extended conjugated chemical structure. The said properties are discussed in detail in the following sections.

4.3 Crystal structures and properties of metal complexes with thio-hydrazone Schiff base ligand (H_2L^3)

4.3.1 Crystal structure of organometallic phenylmercury(II) complex (3)

The organometallic phenylmercury(II) compound **3** crystallized in triclinic form to give a tricoordinated T-shaped geometry (**Figure 15**). Here the monobasic ligand $[(HL^3)^-]$ coordinated the central mercury in a bidentate manner through S and N (N of hydrazone function) and the third coordination of the mercury was satisfied by the C atom of phenyl group of the metal precursor. Thus, by choosing the phenylmercury system, the ligand H_2L^3 has been compelled to act in a bidentate mode leaving another N donor (the oxime N) uncoordinated [35].

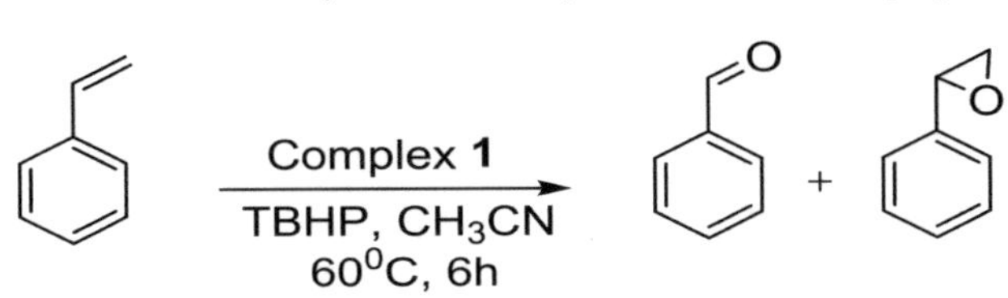

Figure 14.
Oxidation of styrene using $[Ni(L^1)(NCS)_2]$ as catalyst.

The Hg(II) atom remains 0.027(1) Å above the plane. Due to the contribution of electron flow from mercury to the π* orbitals of the phenyl group, the Hg–C bond distance is found shorter than that in the analogous methylmercury(II) compound, where no such electron drifting is observed. The C–S bond gets partial double bond character in the complex, similar to related thiosemicarbazonates of methylmercury (II) and dimethylthallium(III). It is interesting to note that there is no intermolecular π–π interaction between the phenyl rings. But a weak interaction between C(8)–H(8A) and a π group (phenyl ring) links the two phenylmercury molecules into a supramolecular dimer having a C–H π synthon (**Figure 16**) having characteristic H···C_g distance 2.84 Å, where C_g is the midpoint of the phenyl ring.

*4.3.2 Crystal structure of the zinc(II) complex (**4**)*

The X-ray crystal structure shows that due to constrained ligand structure, the [Zn(HL^3)(OAc)(H_2O)].H_2O complex **4** (**Figure 17**) has adopted distorted square pyramidal geometry (τ = 0.05) in which the oxime nitrogen, azomethine nitrogen, deprotonated thiol sulfur of ligand [$(HL^3)^-$], and acetato oxygen defines the equatorial plane, while the axial position is occupied by oxygen (O_4) of coordinated

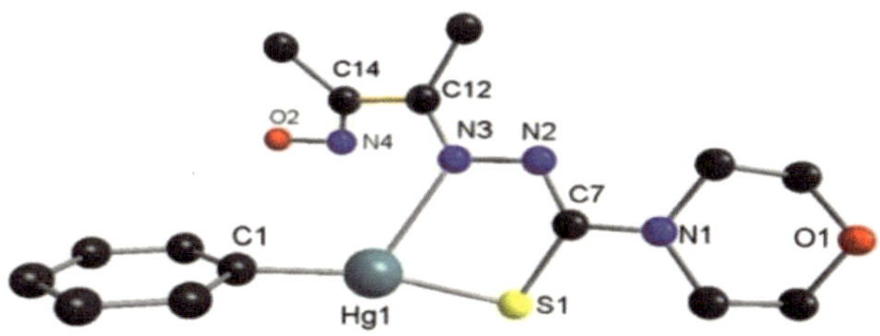

Figure 15.
Perspective view of complex 3 with atom numbering scheme (hydrogen atoms are omitted for clarity).

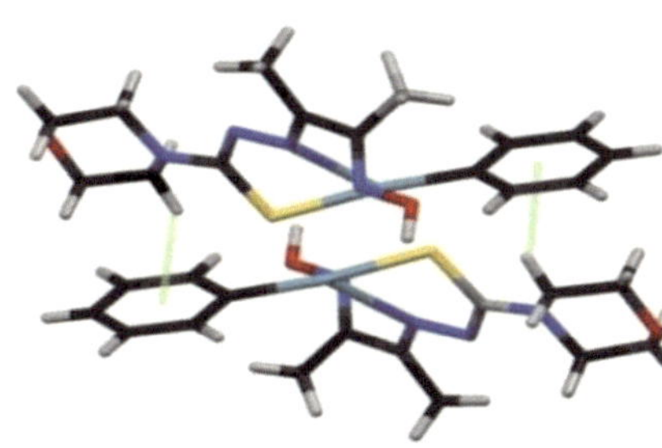

Figure 16.
Weak interaction (C–H···π) between the π-electrons of the phenyl ring with the H atom of ligand in a supramolecular dimer.

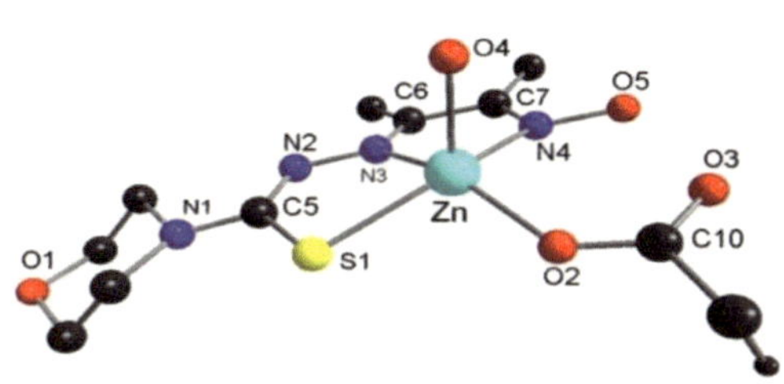

Figure 17.
Perspective view of complex 4 with atom numbering scheme (hydrogen atoms are omitted for clarity).

water molecule. The zinc(II) ion is 0.3766 (8) Å out of the basal plane towards the oxygen (O_4) of coordinated water molecule. The deprotonation of the ligand results in extensive delocalization of charge, and as a consequence the C–S bond length (1.7459(18) Å) is much closer in length to a C–S single bond (1.82 Å), which has partial double bond character [34].

Due to intramolecular hydrogen bonding, the Zn–N(3) (azomethine) distance is slightly shorter than Zn–N(4) (oxime) distance. Here the H(1) of coordinated water molecule is hydrogen bonded to morpholinic oxygen O(1), while H(2) is hydrogen bonded to acetate oxygen O(3), and such H-bonding forms the 1D supramolecular framework diagonal to the *ab* plane (**Figure 18**).

4.3.3 Crystal structure of the cadmium(II) complex (5)

The cadmium(II) complex [Cd(HL^3)$_2$] (**5**) crystallized in a triclinic space group $P\bar{1}$ (2). The central Cd (II) ion is coordinated by two units of monobasic ligand [(HL^3)$^-$] forming trapezoidal bipyramid (**Figure 19**), where the equatorial plane was formed by thiol S1, azomethine N3, oxime N4 of one Schiff base unit, and azomethine N7 atoms of another unit (second) and the axial positions are being coordinated by oxime N8 and thiol S2 atoms of the second Schiff base unit. The C–S bond gets partial double bond character in the complex and is shorter than that in phenylmercury(II) complex (3). The C–N average bond distance confirms double bond character, and deprotonation of the hydrazinic NH proton induces double bond character [35].

Complex **5** forms a 2D supramolecular sheet in the *bc* plane through hydrogen bonding interactions (O–H···O and C–H···O) (**Figure 20**). The average D···A separations and the average D–H···A angle is 145°. These supramolecular sheets stack along (100) the plane. There are no directional weak force interactions (π–π, C–H···π, hydrogen bonding) operating between the two sheets. Therefore, an intermolecular van der Waals interaction might be responsible for such supramolecular stacking. The methyl terminals of the ligand from adjacent sheets face each other in this arrangement.

One of the most interesting parts of the structure is that the crystal structure possesses a remarkably short intermolecular C(sp^3)···O(sp^3) contact [C9···O4* 2.958 (3) Å]. An interesting packing force using an uncommon C(sp^3)···O(sp^3)

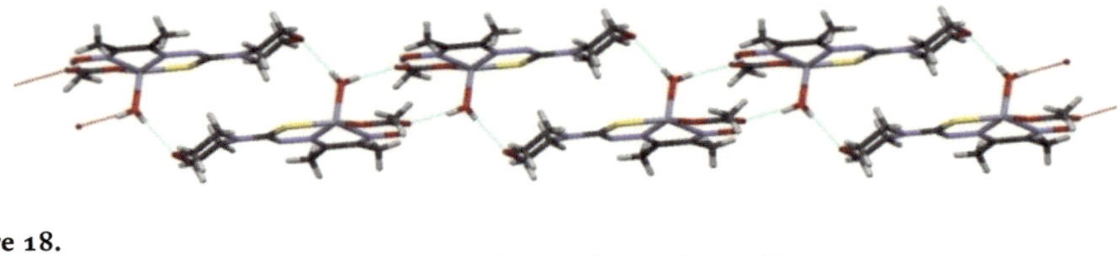

Figure 18.
One-dimensional network through H-bonding in [Zn(HL^3)(OAc)(H_2O)].H_2O.

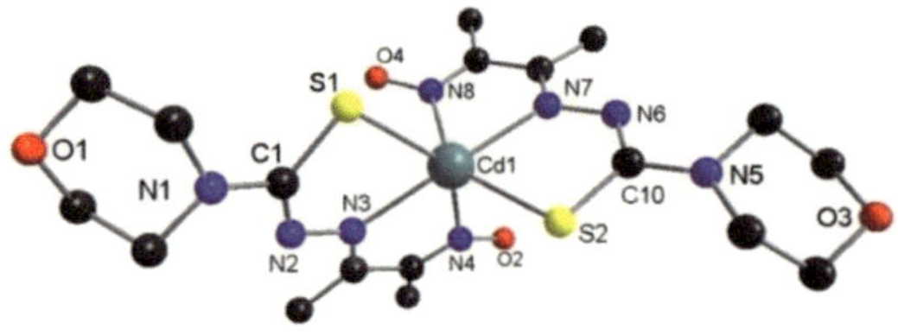

Figure 19.
Perspective view of complex 5 with atom numbering scheme (hydrogen atoms are omitted for clarity).

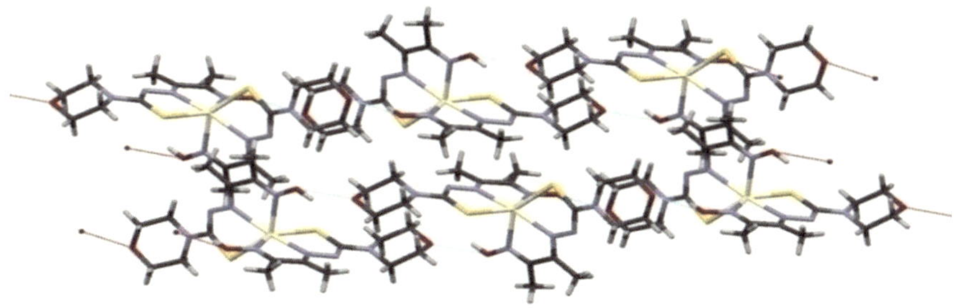

Figure 20.
Two-dimensional network through H-bonding in [Cd(HL^3)$_2$] (5).

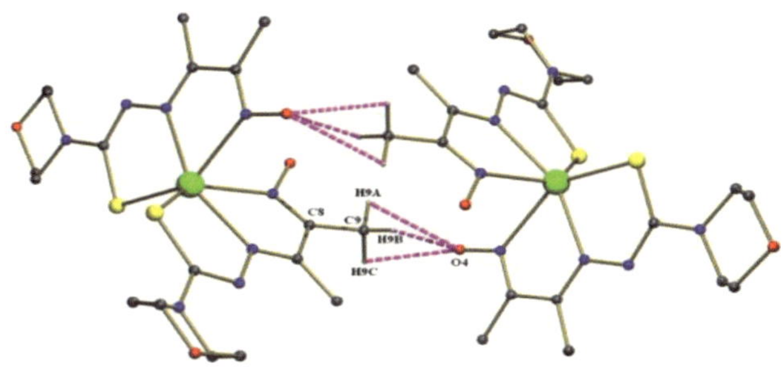

Figure 21.
Trifurcated H-bonding interactions [35] in[Cd(HL^3)$_2$] (5).

interaction has been reported. The three H atoms (H9A, H9B, H9C) of the methyl group of the propylidene moiety form a triangular plane which is supported by the angle ∠C8–C9···O4* of 172.0°. The O4 atom of the oxime is directed towards the center of the plane formed by three H atoms (H9A, H9B, H9C) of the methyl group. The C9–H9A···O4*, C9–H9B···O4*, and C9–H9C···O4* angles are 89.0°, 83.8°, and 97.6°, respectively, which are less than the generally accepted ∠C–H···O angles (110°), and such short bond angles might be responsible for the greater C···O attraction. Thus, the cooperative effect of the three individual interactions acts on the face of the plane of the three methyl hydrogens of the C9 atom due to their high acidic character. This very unusual and nonconventional interaction was termed as a CH_3···O interaction and proved to be a good supramolecular synthon. The trifurcated H-bonding interactions are shown in (**Figure 21**).

*4.3.4 Crystal structure of the chromium(III) complex (**6**)*

The X-Ray single-crystal analysis revealed that complex [Cr(L^3)$_2$]Cl·3H_2O (**6**) forms slightly distorted octahedral geometry with [CrN_4S_2] core having meridional conformation via the oximino N, imine N, and thiol S atoms (**Figure 22**), i.e., the two monobasic tridentate L^3 ligands act as N,N,S donor during formation of the complex [36].

4.3.5 Crystal structure of the iron(III) complex (7)

The X-ray structure of [Fe(L^3)$_2$]Cl·3H_2O (7) is quite similar to that of complex (**6**) with pseudo-octahedral geometry except that the equatorial plane is formed by two oximino N atoms and two thiol S atoms and the axial positions are occupied by two imine N atoms of the ligand (**Figure 23**). The Fe–N(imine) bond lengths are significantly shorter than the Fe–N(oximino) bond lengths because azomethine nitrogen is a stronger base than the oximino nitrogen. The deprotonation of the ligands results in extensive delocalization of charge, and thus the C–S bond gets partial double bond character [36].

The most interesting part of the crystalline structure of complex $[Cr(L^3)_2]Cl{\cdot}3H_2O$ (**6**) and $[Fe(L^3)_2]Cl{\cdot}3H_2O$ (7) is that both the complexes form a 1D supramolecular chain along the ***c***-axis through C–H···O interactions between each molecule (**Figure 24**).

The 1D chains are arranged in parallel direction to form a supramolecular host having channels along the ***c***-axis (**Figure 25**). Such channels are filled by self-assembled "water-chloride" clusters having chair conformation (**Figure 25**).

It is proven by the crystal structure analysis that there are three crystals of water molecules per formula unit of **complexes 6** and 7 along with chloride ion, and these are involved in forming the infinite "water-chloride" cluster [36] with a chair

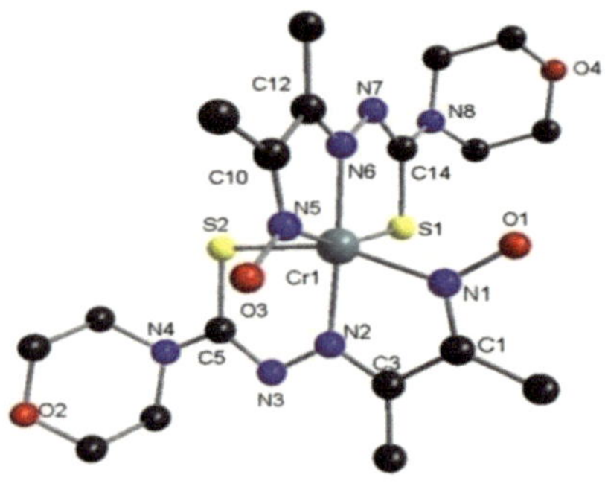

Figure 22.
*Perspective view of complex **6** with atom numbering scheme (hydrogen atoms are omitted for clarity).*

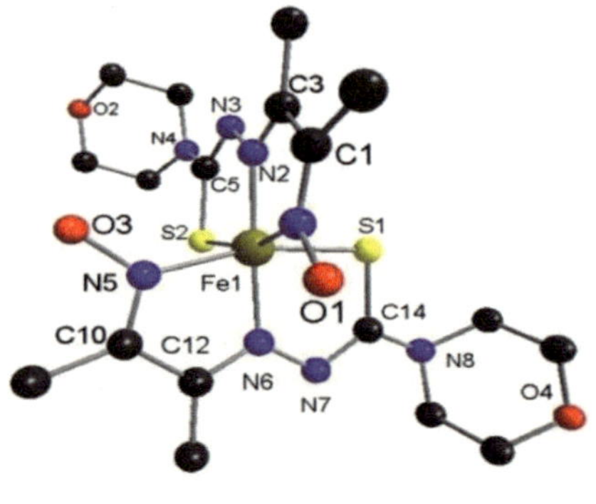

Figure 23.
Perspective view of complex 7 with atom numbering scheme (hydrogen atoms are omitted for clarity).

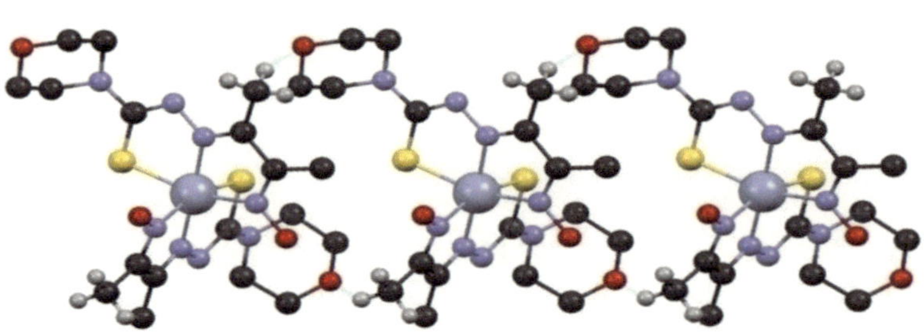

Figure 24.
One-dimensional network through C–H···O interactions in complexes (6) and (7).

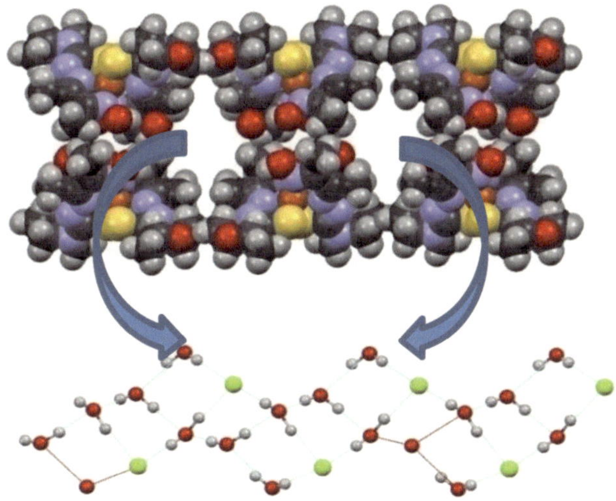

Figure 25.
Supramolecular channels and "water-chloride" cluster along the c-axis in complexes (6) and (7).

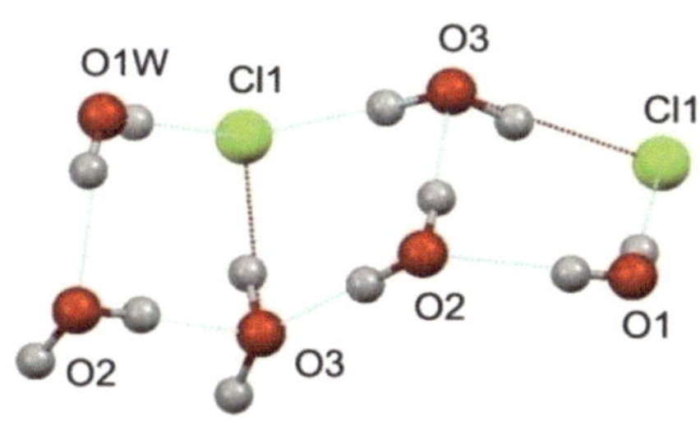

Figure 26.
Chair conformation formed by "water-chloride" cluster in complexes (6) and (7).

conformation. The three water molecules and a chloride molecule form a four-membered cyclic motif, and such three adjacent four-membered rings form the chair conformation (**Figure 26**).

4.3.5.1 Magnetic property and Mössbauer spectroscopy of complex (7)

The variable temperature (2.5–300 K) magnetic moment study shows the temperature dependence of the magnetic susceptibility. The χ_m values at 2.5 and 300 K are 0.33 and 0.005 cm ^{3} mol^{-1}, while the μ_{eff} values are 2.61 and 3.46 B.M., respectively. The detailed study shows that the magnetic moment value consists of a superimposition of both the low-spin and high-spin states. At very low temperature, the 1-D supramolecular species which is formed by strong intermolecular C–H···O interactions and the cooperative interactions with the "water-chloride" cluster between mononuclear spin crossover (SCO) sites stabilize the low-spin state, and thus the high-spin contribution decreases to 21%, and the low-spin contribution increases to 79%. Thus, such variable temperature magnetic behavior may be due to a continuous S = 1/2 to 5/2 spin crossover phenomenon of iron centers [36].

The Mössbauer spectroscopic study also supports that a spin crossover phenomenon exists in the iron(III) complex (7). Both the spin states, low (S = 1/2) and high (S = 5/2), exist in room temperature (300 K) as well as in very low temperature at 20 K. The population density of the electrons decreases from high-spin state to the low-spin state with decreasing temperature (**Table 1**).

Temperature	Spin state	Occupancy
300 K	Low spin (1/2)	49%
	High spin (5/2)	51%
20 K	Low spin (1/2)	77%
	High spin (5/2)	23%

Table 1.
Population density at variable temperature in $[Fe(L^3)_2]Cl\cdot 3H_2O$ (7).

Cell parameters	$[Zn(L^3)_2].2H_2O$ (8)	$[Ni(L^3)(OAc)]$ (9)	$[Co(L^3)(OAc)].H_2O$ (10)	$[Cu(L^3)(OAc)].H_2O$ (11)
System	Triclinic	Monoclinic	Monoclinic	Monoclinic
V (Å^3)	1294.88	810.3	1127.7	1044.17
a (Å)	10.297368	6.364172	19.600876	18.953438
b (Å)	11.32531	27.497931	5.53422	6.365518
c (Å)	12.345947	4.686936	12.32786	8.729238
α	111.516869	90	90	90
β	103.288712	98.92	122.51	97.5
γ	91.155464	90	90	90

Table 2.
Powder X-ray diffraction data.

4.3.6 PXRD, SEM, and EDX studies of complexes (8)–(11)

Suitable single crystals of complexes (**8**)–(**11**) could not be grown even after repeated efforts; hence they were characterized by the powder X-ray diffraction study of the compounds [28, 29]. Some of the important lattice parameters of the PXRD study are summarized in **Table 2**.

The SEM investigation of all the above complexes, the ground powders, and the fracture surfaces indicates that the grain size distribution is not uniform, and submicron grains (finely ground powder) as well as grains (fracture surfaces) even above 20 μm (for complex 8) and above 10 μm (for complexes 9–11) have been observed [37, 38].

The formation of metal–ligand complexes and the presence of metal along with C and S within the metal complexes have been substantiated by the EDX analysis.

4.4 Electrical conductivity

To explore the utility of the metal complexes as functional materials, the electrical conductivity study was performed, and it shows the semiconducting nature of the complexes [33, 37, 38].

The samples for the measurement of electrical conductivity were prepared from the complexes in the form of tablets of approximately thickness ~0.1 cm at a pressure of ca. 1×10^8 Pascal. These tablets were placed between two copper electrodes covered with silver paste, and contacts of the prepared tablets were to be Ohmic or not. A two-probe method was used to investigate the electrical conductivities of the complex tablets by measuring the current through the probes with a high impedance electrometer (Keithley 6514) upon application of a DC voltage current supplied by a programmable source of voltage (Keithley 230). The conductivities were calculated by using the general equation of $\sigma = (I/V_c)(d/a)$, where (I) is the current in ampere, V_c the potential drop across the sample of cross-sectional area (a), and is the thickness (d).

Variation of electrical conductivity of a compound behaving like semiconductor with temperature can be obtained by the Arrhenius equation:

$$\sigma = \sigma_o \exp\left(-E_a/kT\right) \tag{1}$$

where σ is the electrical conductivity, σ_o denotes the pre-exponential factor, E_a is the activation energy for this thermally activated process, and k is the Boltzmann constant. Generally, "log σ" is plotted against "1000/T" which is expected to have a linearly fitted graph. The thermal energy of activation (E_a) is determined from the slope of the graph.

If the graph obtained is linear (i.e., fitted with one straight line), then it may be concluded that no molecular rearrangement occurred during heating and the compound will have only one E_a. In this case the electrical conduction is mainly due to the intrinsic conducting property, whereas the nonlinear plot clearly indicates molecular rearrangement during heating of the compound. In case of nonlinear plot, more than one straight line are fitted against the linear parts of the plot. One straight line corresponding to lower temperature range is known as region I, and the other straight line corresponding to the higher temperature range is known as region II. In this case the compound will have two distinct energy of activations (E_{a1} and E_{a2}).

Conduction corresponding to the region I is attributed to the intermolecular conduction via weak force interactions between the molecules. The charge carriers hop near Fermi level within the localized state. Delocalized π-electrons are mainly responsible for this conduction, whereas conduction corresponding to the region II is attributed to intramolecular conduction between the metal center and the ligand center within a metal complex. This conduction occurs due to tunneling of electrons between equivalent HOMO and LUMO of the ligand and metal ion, respectively. Such tunneling of electrons through the intermolecular potential barrier is reinforced through π–π stacking and extensive H-bonding [24]. Depending on the availability of π-electrons, the compound behaves like n-type semiconductor.

From the Arrhenius plots (**Figure 27**), the electronic parameters, i.e., activation energy of electrical conduction (E_a) and the energy gap for directly allowed transitions of metal complexes (2) and (**8–11**), are calculated, and the results obtained are summarized in **Table 3**.

It is also very clear from the Arrhenius plots that the conductivity of metal complexes generally increases with increase in temperature. At room temperature

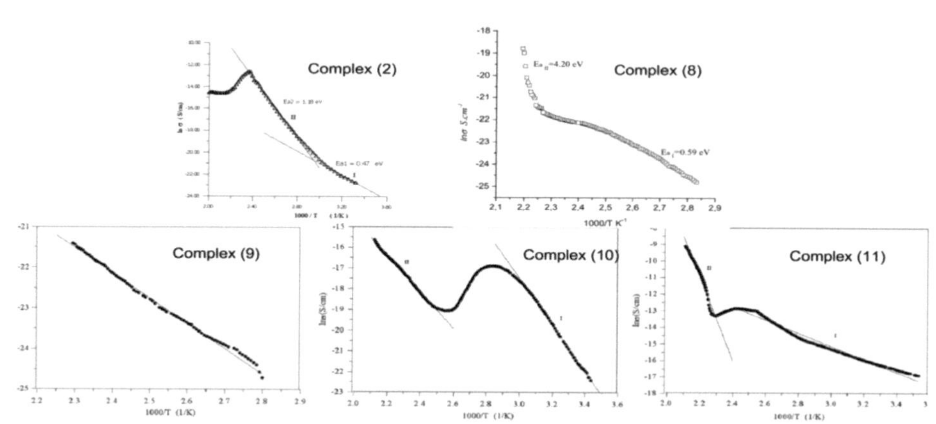

Figure 27.
Temperature dependence. Electrical conductivity curves of the complexes [24, 28, 29].

Complex	E_{a1} (eV)	E_{a2} (eV)	E_{gd} (eV)
	(Lower temp)	(Higher temp)	
$[VOL^2]$ (**2**)	0.48	1.18	3.45
$[Zn(L^3)_2].2H_2O$ (**8**)	0.59	4.2	2.52
$[Ni(L^3)(OAc)]$ (**9**)	0.54	—	2.75
$[Co(L^3)(OAc)].H_2O$ (**10**)	0.97	0.76	2.37
$[Cu(L^3)(OAc)].H_2O$ (**11**)	0.34	2.14	1.58

Table 3.
Activation energies and direct band gap values.

they behave as an insulator, while at higher temperature the semiconducting nature of complexes is observed.

4.5 Optical properties

Optical absorption spectra was taken by using a UV–VIS spectrophotometer (Perkin Elmer Lambda 2S/45 Double Beam) and measured as function of wavelength in the wavelength range 190–1100 nm.

The energy band gaps and the nature of the optical transitions involved in the metal complex framework systems have been practically determined by the fundamental absorption edge analysis of the recorded optical transitions using the theory of Mott and Davis [39]. It is also observed that the semiconducting behavior of a material increases with rise in temperature which may also damage the actual molecular structure of the material. Hence, Tauc method is used to calculate the energy band gap through optical absorption properties [40].

Utilizing the relation between the optical linear absorption coefficient (α) with photon energy (hν), the energy band gap (E_g) between the top of the valence band and bottom of the conduction band can be determined using equation (Eq. (2)):

$$\alpha h\nu = A\left(h\nu - E_g\right)^n \tag{2}$$

where A is a constant characteristic parameter of the respective transition independent of ν.

The values of n depend on the kind of optical transitions. For directly allowed, directly forbidden, indirectly allowed, and indirectly forbidden transitions, the values of n are ½, 3/2, 2, and 3, respectively. Thus the energy band gap for directly allowed (Egd) and indirectly allowed (Egi) transitions can be determined by relating Eq. (2) as follows:

$$\alpha h\nu = A_d\left(h\nu - E_{gd}\right)^{1/2} \tag{3}$$

and

$$\alpha h\nu = A_i\left(h\nu - E_{gi}\right)^2 \tag{4}$$

where E_{gd} and E_{gi} are direct and indirect energy gaps, respectively.

To calculate the direct and indirect energy band gap, we need to plot a curve of $(\alpha h\nu)^2$ against f(hν) and $(\alpha h\nu)^{1/2}$ against f(hν) and then by the extrapolation of the most linear part of the curve to zero.

The satisfactory graphs were obtained for the metal complexes (**2**) and (**8**)–(**11**) by plotting $(\alpha h\nu)^2$ against f(hν). Therefore, the energy gaps determined correspond to the direct energy gap (E_{gd}). The E_{gd} values of the metal complexes are collected in **Table 3**. The comparison of the band gap cannot be directly related to the atomic number of the metals in the complexes.

5. Conclusion

In this review, the synthesis, crystal structure, and solid-state properties of three Schiff base ligands derived from diacetylmonoxime with diethylenetriamine, 1,3-diaminopropane-2-ol, and morpholine N-thiohydrazide and their metal complexes have been vividly discussed. A zwitterionic nitrogen–sulfur heterocyclic compound with nonbonded S···S interaction has also been reported to be formed by the reaction of diacetylmonoxime with morpholine N-thiohydrazide under long refluxing (16 h) condition in ethanol. The single X-ray crystal structures have shown many beautiful weak force interactions including a CH_3···O trifurcated interface communication. Wherever the single-crystal structures could not be grown, the PXRD study has enlightened their structural features. The electrical and optical properties also explored the semiconducting nature of some of the metal complexes. It is also observed that the electron transport process gets influenced by the supramolecular frameworks of the metal complexes.

Acknowledgements

One of the authors (S.S.) is thankful to the UGC (ERO), Kolkata, for financial grants (MRP) to carry out a part of this work and also to Prof. Y. Aydogdu, Department of Physics, Gazi University, and Dr. S. Biswas, our lab-mate for some useful discussion.

Author details

Palash Mandal[1,2], Uttam Das[3], Kamalendu Dey[1†] and Saikat Sarkar[4*]

1 Department of Chemistry, University of Kalyani, Kalyani, West Bengal, India

2 Lalgopal High (H.S.) School, Ranaghat, West Bengal, India

3 Department of Chemistry, Kalyani Government Engineering College, Kalyani, West Bengal, India

4 Department of Chemistry, Chakdaha College, Chakdaha, West Bengal, India

*Address all correspondence to: saikat_s@rediffmail.com

† Prof. Kamalendu Dey expired on September 01, 2019.

References

[1] Yoon TP, Jacobsen EN. Privileged chiral catalysts. Science. 2003;**299**: 1691-1693. DOI: 10.1126/science.1083622

[2] Slassi S, Aarjane M, Yamni K, Amine A. Synthesis, crystal structure, DFT calculations, Hirshfeld surfaces, and antibacterial activities of Schiff base based on imidazole. Journal of Molecular Structure. 2019;**1197**: 547-554. DOI: 10.1016/j.molstruc.2019.07.071

[3] Qian HY, Sun N. Synthesis and crystal structures of manganese (III) complexes derived from bis-Schiff bases with antibacterial activity. Transition Metal Chemistry. 2019;**44**:501-506. DOI: 10.1007/s11243-018-00296-x

[4] Sarkar S, Dey K. A series of transition and non-transition metal complexes from a N_4O_2 hexadentate Schiff base ligand: Synthesis, spectroscopic characterization and efficient antimicrobial activities. Spectrochimica Acta Part A. 2010;**77**:740-748. DOI: 10.1016/j.saa.2010.06.041

[5] Sharma M, Chauhan K, Srivastava RK, Singh SV, Srivastava K, Saxena JK, et al. Design and synthesis of a new class of 4-aminoquinolinyl-and 9-anilinoacridinyl Schiff base hydrazones as potent antimalarial agents. Chemical Biology & Drug Design. 2014;**84**: 175-181. DOI: 10.1111/cbdd.12289

[6] Ziegler J, Schuerle T, Pasierb L, Kelly C, Elamin A, Cole KA, et al. The propionate of heme binds N_4O_2 Schiff base antimalarial drug complexes. Inorganic Chemistry. 2000;**39**: 3731-3733. DOI: 10.1021/ic000295h

[7] Chang EL, Simmers C, Knight DA. Cobalt complexes as antiviral and antibacterial agents. Pharmaceuticals. 2010;**3**:1711-1728. DOI: 10.3390/ph3061711

[8] Chen Y, Li P, Su S, Chen M, He J, Liu L, et al. Synthesis and antibacterial and antiviral activities of myricetin derivatives containing a 1, 2, 4-triazole Schiff base. RSC Advances. 2019;**9**: 23045-23052. DOI: 10.1039/C9RA05139B

[9] Venkateswarlu K, Ganji N, Daravath S, Kanneboina K, Rangan K. Crystal structure, DNA interactions, antioxidant and antitumor activity of thermally stable Cu (II), Ni (II) and Co (III) complexes of an N, O donor Schiff base ligand. Polyhedron. 2019;**171**:86-97. DOI: 10.1016/j.poly.2019.06.048

[10] da Silveira VC, Luz JS, Oliveira CC, Graziani I, Ciriolo MR, da Costa Ferreira AM. Double-strand DNA cleavage induced by oxindole-Schiff base copper (II) complexes with potential antitumor activity. Journal of Inorganic Biochemistry. 2008;**102**: 1090-1103. DOI: 10.1016/j.jinorgbio.2007.12.033

[11] Das U, Pattanayak P, Santra MK, Chattopadhyay S. Synthesis of new oxido-vanadium complexes: Catalytic properties and cytotoxicity. Journal of Chemical Research. 2018;**42**:57-62. DOI: 10.3184/174751918X15168821806597

[12] Cozzi PG. Metal–Salen Schiff base complexes in catalysis: Practical aspects. Chemical Society Reviews. 2004;**33**: 410-421. DOI: 10.1039/B307853C

[13] Gupta K, Sutar AK. Catalytic activities of Schiff base transition metal complexes. Coordination Chemistry Reviews. 2008;**252**:1420-1450. DOI: 10.1016/j.ccr.2007.09.005

[14] Pouralimardan O, Chamayou AC, Janiak C, Hosseini-Monfared H. Hydrazone Schiff base-manganese (II) complexes: Synthesis, crystal structure and catalytic reactivity. Inorganica

Chimica Acta. 2007;**360**:1599-1608. DOI: 10.1016/j.ica.2006.08.056

[15] Bagherzadeh M, Mahmoudi H, Ataie S, Hafezi-Kahnamouei M, Shahrokhian S, Bellachioma G, et al. Synthesis, characterization, and comparison of two new copper (II) complexes containing Schiff-base and diazo ligands as new catalysts in CuAAC reaction. Inorganica Chimica Acta. 2019;**492**:213-220. DOI: 10.1016/j.ica.2019.04.036

[16] Mondal P, Parua SP, Pattanayak P, Das U, Chattopadhyay S. Synthesis and structure of copper (II) complexes: Potential cyanide sensor and oxidase model. Journal of Chemical Sciences. 2016;**128**:803-813. DOI: 10.1007/s12039-016-1063-7

[17] Ghosh K, Banerjee A, Bauzá A, Frontera A, Chattopadhyay S. One pot synthesis of two cobalt (III) Schiff base complexes with chelating pyridyltetrazolate and exploration of their bio-relevant catalytic activities. RSC Advances. 2018;**8**:28216-28237. DOI: 10.1039/C8RA03035A

[18] Sedighipoor M, Kianfar AH, Mohammadnezhad G, Görls H, Plass W. Unsymmetrical palladium (II) N, N, O, O-Schiff base complexes: Efficient catalysts for Suzuki coupling reactions. Inorganica Chimica Acta. 2018;**476**:20-26. DOI: 10.1016/j.ica.2018.02.007

[19] Andruh M, Branzea DG, Gheorghe R, Madalan AM. Crystal engineering of hybrid inorganic–organic systems based upon complexes with dissymmetric compartmental ligands. CrystEngComm. 2009;**11**:2571-2584. DOI: 10.1039/B909476H

[20] Ganguly R, Sreenivasulu B, Vittal JJ. Amino acid-containing reduced Schiff bases as the building blocks for metallasupramolecular structures. Coordination Chemistry Reviews. 2008;**252**:1027-1050. DOI: 10.1016/j.ccr.2008.01.005

[21] Liu X, Hamon JR. Recent developments in penta-, hexa-and heptadentate Schiff base ligands and their metal complexes. Coordination Chemistry Reviews. 2019;**389**:94-118. DOI: 10.1016/j.ccr.2019.03.010

[22] Furukawa H, Cordova KE, O'Keeffe M, Yaghi OM. The chemistry and applications of metal-organic frameworks. Science. 2013;**341**: 1230444(1-12). DOI: 10.1126/science.1230444

[23] Li C, Tang H, Fang Y, Xiao Z, Wang K, Wu X, et al. Bottom-up assembly of a highly efficient metal–organic framework for cooperative catalysis. Inorganic Chemistry. 2018;**57**: 13912-13919. DOI: 10.1021/acs.inorgchem.8b02434

[24] Xiong G, Chen XL, You LX, Ren BY, Ding F, Dragutan I, et al. La-metal-organic framework incorporating Fe_3O_4 nanoparticles, post-synthetically modified with Schiff base and Pd. A highly active, magnetically recoverable, recyclable catalyst for CC cross-couplings at low Pd loadings. Journal of Catalysis. 2018;**361**:116-125. DOI: 10.1016/j.jcat.2018.02.026

[25] Ross TM, Neville SM, Innes DS, Turner DR, Moubaraki B, Murray KS. Spin crossover in iron (III) Schiff-base 1-D chain complexes. Dalton Transactions. 2010;**39**:149-159. DOI: 10.1039/B913234A

[26] Yang H, Liu SS, Meng YS, Zhang YQ, Pu L, Yu XQ. Magnetic properties and theoretical calculations of mononuclear lanthanide complexes with a Schiff base coordinated to Ln (III) ion in a monodentate coordination mode. Inorganica Chimica Acta. 2019; **494**:8-12. DOI: 10.1016/j.ica.2019.04.051

[27] Miyasaka H, Clérac R, Wernsdorfer W, Lecren L, Bonhomme C, Sugiura KI, et al. A dimeric manganese (III) tetradentate Schiff base complex as a single-molecule magnet. Angewandte Chemie International Edition. 2004;**43**: 2801-2805. DOI: 10.1002/anie.200353563

[28] Nandy M, Shit S, Rosair G, Gómez-García C. Synthesis, characterization and magnetic studies of a tetranuclear manganese (II/IV) compound incorporating an amino-alcohol derived Schiff base. Magnetochemistry. 2018;**4**: 57-67. DOI: 10.3390/magnetochemistry4040057

[29] Sarkar S, Nag SK, Chattopadhyay AP, Dey K, Islam SM, Sarkar A, et al. Synthesis, structure and catalytic activities of nickel (II) complexes bearing N_4 tetradentate Schiff base ligand. Journal of Molecular Structure. 2018;**1160**:9-19. DOI: 10.1016/j.molstruc.2018.01.035

[30] Sarkar S, Biswas S, Liao MS, Kar T, Aydogdu Y, Dagdelen F, et al. An attempt towards coordination supramolecularity from Mn (II), Ni (II) and Cd (II) with a new hexadentate [N_4O_2] symmetrical Schiff base ligand: Syntheses, crystal structures, electrical conductivity and optical properties. Polyhedron. 2008;**27**:3359-3370. DOI: 10.1016/j.poly.2008.07.034

[31] Costes JP, Duhayon C, Vendier L, Mota AJ. Reactions of a series of ZnL, CuL and NiL Schiff base and non-Schiff base complexes with MCl_2 salts (M = Cu, Ni, Mn): Syntheses, structures, magnetic properties and DFT calculations. New Journal of Chemistry. 2018;**42**:3683-3691. DOI: 10.1039/C7NJ04347C

[32] Majumdar D, Das D, Sreejith SS, Das S, Biswas JK, Mondal M, et al. Dicyanamide-interlaced assembly of Zn (II)-Schiff-base complexes derived from salicylaldimino type compartmental ligands: Syntheses, crystal structures, FMO, ESP, TD-DFT, fluorescence lifetime, in vitro antibacterial and anti-biofilm properties. Inorganica Chimica Acta. 2019;**489**:244-254. DOI: 10.1016/j.ica.2019.02.022

[33] Sarkar S, Aydogdu Y, Dagdelen F, Bhaumik BB, Dey K. X-ray diffraction studies, thermal, electrical and optical properties of oxovanadium (IV) complexes with quadridentate Schiff bases. Materials Chemistry and Physics. 2004;**88**:357-363. DOI: 10.1016/j.matchemphys.2004.08.001

[34] Biswas S, Yap GP, Dey K. Reaction of diacetylmonoxime with morpholine N-thiohydrazide in the absence and in presence of a metal ion: Facile synthesis of a thiadiazole derivative with non-bonded S···S interaction. Polyhedron. 2009;**28**:3094-3100. DOI: 10.1016/j.poly.2009.06.091

[35] Biswas S, Sarkar S, Steele IM, Sarkar S, Mostafa G, Bhaumik BB, et al. Two-dimensional supramolecular assembly of phenylmercury (II) and cadmium (II) complexes with a tridentate thiohydrazone NNS donor ligand: Synthesis, coordination behavior and crystal structure. Polyhedron. 2007; **26**:5061-5068. DOI: 10.1016/j.poly.2007.07.027

[36] Saha R, Biswas S, Steele IM, Dey K, Mostafa G. A supramolecular spin crossover Fe (III) complex and its Cr (III) isomer: Stabilization of water–chloride cluster within supramolecular host. Dalton Transactions. 2011;**40**: 3166-3175. DOI: 10.1039/C0DT01256D

[37] Dagdelen F, Aydogdu Y, Dey K, Biswas S. Synthesis, characterization and solid-state properties of [Zn(Hdmmthiol)$_2$].2H_2O complex. The European Physical Journal Plus. 2016; **131**:143-150. DOI: 10.1140/epjp/i2016-16143-2

[38] Biswas S, Dagdelen F, Aydogdu Y, Dey K. Structural, electrical and optical properties of metal complexes of NNS donor ligand. Materials Chemistry and Physics. 2011;**129**:1121-1125. DOI: 10.1016/j.matchemphys.2011.05.071

[39] Mott NF, Davis EA. Electronic Processes in Non-crystalline Materials. Oxford: Clarendon Press; 1971. DOI: 10.1002/crat.19720070420

[40] Tauc J, Grigorovici R, Vancu A. Optical properties and electronic structure of amorphous germanium. Physica Status Solidi B. 1966;**15**:627-637. DOI: 10.1002/pssb.19660150224

Chapter 2

Stability of Metal Complexes

Senthilkumar Muthaiah, Anita Bhatia
and Muthukumar Kannan

Abstract

The stability of coordination complex is an important factor that decides the stability and reactivity of a metal complex. The stability of metal complex is governed by two different aspects such as thermodynamic and kinetic stabilities. The correlation between stability and reactivity of coordination compounds has been described in this chapter. This chapter also enlists the factors influencing the stability of metal complexes such as the nature of metal ions, ligands, bonding between metal ions and ligands, etc. In addition, the methods available for the determination of stability constants are given in detail.

Keywords: stability, reactivity, coordination complexes, stability constants

1. Introduction

The stability of metal complex generally means that it exists under favorable conditions without undergoing decomposition and has a considerable shelf life period [1]. The term stability of metal complex cannot be generalized since the complex may be stable to one reagent/condition and may decompose in presence of another reagent/condition. The stability of metal complexes can be explained with the help of two different aspects, namely, thermodynamic stability and kinetic stability [2]. Nevertheless, a metal complex is said to be stable if it does not react with water, which would lead to a decrease in the free energy of the system, i.e., thermodynamic stability. On the other hand, the complex is said to possess kinetic stability if it reacts with water to form a stable product and there is a known mechanism through which the reaction can proceed. For example, the system may not have sufficient energy available to break a strong bond, although once the existing bond is broken it could be replaced by new bond which is stronger than the older one [1]. Stability of complex compound is assigned to be its existence in aqueous solution with respect to its bond dissociation energy, Gibbs free energy, standard electrode potential, pH of the solution, and rate constant or activation energy for substitution reactions.

1.1 Thermodynamic stability

Thermodynamic stability of a complex refers to its tendency to exist under equilibrium conditions. It determines the extent to which the complex will be formed or be converted into another complex at the point of equilibrium. In other words, thermodynamic stability of complexes is the measure of tendency of a metal ion to selectively form a specific metal complex and is directly related to the metal-ligand bond energies. The thermodynamic stability of complexes is represented by formation constant. The formation constant is also known as stability constant, which is the equilibrium constant obtained for the formation metal complex [1, 2].

In general, the metal complexes are not prepared from their corresponding starting materials in gaseous phase but are prepared in aqueous solution. In aqueous solution, a metal cation gets hydrated to give aqua complex of the type $[M(H_2O)_x]^{n+}$. When a ligand replaces water molecule from aqua complex ion, a new metal complex is formed and equilibrium is established as shown:

$$[M(H_2O)_x]^{n+} + L \xrightleftharpoons{K_f} [M(H_2O)_{x-1}L]^{n+} + H_2O \quad (1)$$

where x is the number of water molecules, n is the oxidation number of the metal cation, and L is the neutral and monodentate ligand. For simplicity, the above reaction can be written in generalized form as given:

$$M + L \xrightleftharpoons{K_f} ML \quad (2)$$

The equilibrium constant K_f of the reaction is given by:

$$K_f = \frac{[ML]}{[M][L]} \quad (3)$$

In the above equation, the concentration of water is not included. Since the solution is dilute, the water molecules which enter the bulk solution do not have much influence on the equilibrium constant. It is observed from Eq. (3) that the higher the value of K_f, the greater will be the stability of the complex formed. A high value of the equilibrium constant ($K_f > 1.0$) also indicates that at equilibrium the activity of complex ML is larger than the product of activities of M and L. Thus, large value of K_f indicates that the ligand L binds to the metal ion more strongly than H_2O and hence L is a stronger ligand than H_2O. If K_f is less than 1.0, then ligand L is weaker than H_2O. Thus stability constant is used as a measure of thermodynamic stability of the complex. With a few exceptions, the value of successive stability constants decreases regularly from K_1 to K_n, that is, $K_1 > K_2 > K_3 > \ldots > K_{n-1} > K_n$. This trend is illustrated by taking formation of $[Cd(NH_3)_4]^{2+}$ as an example [3, 4]:

$$Cd^{2+} + NH_3 \rightleftharpoons [Cd(NH_3)]^{2+}\ ;\quad K_1 = 10^{2.65}$$

$$[Cd(NH_3)]^{2+} + NH_3 \rightleftharpoons [Cd(NH_3)_2]^{2+}\ ;\quad K_2 = 10^{2.10}$$

$$[Cd(NH_3)_2]^{2+} + NH_3 \rightleftharpoons [Cd(NH_3)_3]^{2+}\ ;\quad K_3 = 10^{1.44}$$

$$[Cd(NH_3)_3]^{2+} + NH_3 \rightleftharpoons [Cd(NH_3)_4]^{2+}\ ;\quad K_4 = 10^{0.93}$$

The steady decrease in the value of stepwise formation constants from K_1 to K_n is due to:

i. Increase in the number of ligands in coordination sphere that causes to decrease the number of H_2O molecules to be replaced and thus the probability of replacement of water molecules decreased

ii. Electrostatic factor

iii. Steric hindrance with increase in number of ligands

iv. Statistical factors (number of replaceable positions)

However, in some cases, it is found that $K_{n+1} > K_n$ because of unusual structural changes and changes in electronic configuration of the metal ion. The change in electronic structure of the metal ion causes the variation in the crystal field stabilization energy (CFSE). The complex with higher CFSE value will be stable, and the equilibrium constant for that complex formation will be high. One such example is the formation of $[CdBr_4]^{2-}$ complex in aqueous solution. The reaction of aqua complex $[Cd(H_2O)_6]^{2+}$ with Br^- ligand exhibits four stepwise equilibrium or stepwise formation constants K_1, K_2, K_3, and K_4. The order of stepwise formation constants is observed as follows, $K_1 > K_2 > K_3 < K_4$, which is not in agreement with the common trend of $K_1 > K_2 > K_3 > K_4$. Aqua complex of most of the M^{2+} ions including Cd^{2+} are octahedral, whereas the halo complexes of Cd^{2+} ion are tetrahedral. The reaction sequence for the formation of $[CdBr_4]^{2-}$ is given as follows:

$$[Cd(H_2O)_6]^{2+} + Br^- \underset{}{\overset{K_1}{\rightleftharpoons}} [Cd(H_2O)_5Br]^+ + H_2O$$

$$[Cd(H_2O)_5Br]^+ + Br^- \overset{K_2}{\rightleftharpoons} [Cd(H_2O)_4Br_2] + H_2O$$

$$[Cd(H_2O)_4Br_2] + Br^- \overset{K_3}{\rightleftharpoons} [Cd(H_2O)_3Br_3]^- + H_2O$$

$$[Cd(H_2O)_3Br_3]^- + Br^- \overset{K_4}{\rightleftharpoons} [CdBr_4]^{2-} + 3H_2O$$

In the final step, there is an unusual structural change from six coordinated octahedral $[Cd(H_2O)_3Br_3]^-$ complex to four coordinated tetrahedral $[CdBr_4]^{2-}$ complex in addition to change in the electronic configuration which lead to $K_4 > K_3$.

The formation constant (K_f) is related to the standard Gibbs free energy change ($\Delta G°$) and standard electrode potential ($E°$) according to following equations:

$$\Delta G° = -RT\ln K_f \tag{4}$$

$$\Delta G° = -nFE° \tag{5}$$

$$\text{hence } RT\ln K_f = nFE° \tag{6}$$

Since $\Delta G°$ is a thermodynamic property, the formation constant is the measure of thermodynamic stability. From Eqs. (4) to (6), it can be interpreted that the thermodynamic stability of a complex can be measured in terms of formation constant, Gibbs free energy change, and standard electrode potential. A high negative value of $\Delta G°$ indicates that the position of equilibrium favors the product (complex); hence the complex formed will be more stable.

1.2 Stepwise formation of complex and stepwise formation constants

The formation constant describes the formation of a complex from metal cation and ligands. Bjerrum (1941) defined that the formation of a metal complex in aqueous solution takes place by replacing the water molecule by another ligand (L)

[5, 6]. It is assumed that this reaction does not occur in a single step but occurs in several steps, and each step is characterized by its individual equilibrium constant called as stepwise formation constant (K). For example, consider the formation of a complex $[ML_n]$ formed by the following reactions:

$$M + L \overset{K_1}{\rightleftharpoons} ML$$

By assuming the value of activity coefficients as unity, the equilibrium constant K_1 for the complex (ML) having one ligand (L) will be given as

$$K_1 = \frac{[ML]}{[M][L]}$$

When the metal complex ML reacts with one more ligand L,

$$ML + L \overset{K_2}{\rightleftharpoons} ML_2$$

and the equilibrium constant K_2 will be

$$K_2 = \frac{[ML_2]}{[ML][L]}$$

Similarly, for the formation of the complex ML_n from ML_{n-1} and L, the equilibrium constant is represented as follows,

$$K_n = \frac{[ML_n]}{[ML_{n-1}]\,[L]}$$

The equilibrium constants K_1, K_2, ..., K_n are known as stepwise formation constants. On the other hand, the equilibrium constant for the overall reaction may be considered as

$$M + L \overset{\beta_1}{\rightleftharpoons} ML, \qquad \beta_1 = \frac{[ML]}{[M][L]}$$

$$M + 2L \overset{\beta_2}{\rightleftharpoons} ML_2, \qquad \beta_2 = \frac{[ML_2]}{[M][L]^2}$$

$$M + 3L \overset{\beta_3}{\rightleftharpoons} ML_3, \qquad \beta_3 = \frac{[ML_3]}{[M][L]^3}$$

$$M + nL \overset{\beta_n}{\rightleftharpoons} ML_n, \qquad \beta_n = \frac{[ML_n]}{[M][L]^n}$$

where β_1, β_2, β_3, ..., β_n are the equilibrium constants called as overall formation constants and K_1, K_2, K_3, ..., K_n are stepwise stability or formation constants. The products of stepwise constants are Ks and βs are related one another. For example, consider the product of stepwise formation constants K_1, K_2, K_3, ..., K_n.

$$K_1 \times K_2 \times K_3 \times \ldots\ldots K_n = \frac{[ML]}{[M][L]} \times \frac{[ML_2]}{[ML][L]} \times \frac{[ML_3]}{[ML_2][L]} \times \cdots\cdots \frac{[MLn]}{[ML_{n-1}][L]}$$
$$= \frac{[ML_n]}{[M][L]^n} = \beta$$

Above equation indicates that the overall formation constant (β) is equal to the product of the stepwise formation constant K_1, K_2, K_3, … , K_n.

1.3 Kinetic stability

Kinetic stability is related to the reactivity of the metal complexes in solution and deals with the rate of the reaction, its activation energy, etc. Kinetic stability is also related to how fast a compound reacts rather than how stable it is. It aids in determining the rate at which the reaction occurs to establish the equilibrium [7].

The term kinetic stability of complexes is classified into labile and inert by Taube on the basis of rate of the reactions. When the rate of substitution of ligands is high, the complex is said to be labile. For example, the copper complex of the formula $[Cu(NH_3)_4(H_2O)_2]^{2+}$ is labile. In aqueous solution the complex is blue in color, and when concentrated hydrochloric acid is added to this solution, the solution turns green giving rise to complex $[CuCl_4]^{2+}$. On the other hand, in inert complexes the rate of ligand exchange is very slow, and the ligands are very exchanged with difficulty. For example, the cobalt complex $[Co(NH_3)_6]^{3+}$ reacts slowly, and no reaction takes place at room temperature when conc. HCl was added to the aqueous solution. However, only one NH_3 ligand was found to be substituted by Cl ligand, when the aqueous solution of the complex was heated with 6M hydrochloric acid.

2. Relation between thermodynamic and kinetic stabilities

For metal complexes, the stability and reactivity are described in thermodynamic and kinetic terms, respectively. In particular, the terms stable and unstable are related to thermodynamic aspects, whereas labile and inert terms are related to kinetic aspects. As a rule of thumb, a metal complex is said to be labile if it reacts within 1 min at 25°C, and if it takes longer time, it is considered to be inert.

Thermodynamic stability refers to the energy change that occurs while starting materials are converted to products, that is, ΔG, for the reaction. The change in free energy is given by the equation $\Delta G = \Delta H - T\Delta S = -RT\ln K$, where ΔS is the entropy, ΔH is the enthalpy, and K is the equilibrium constant for the reaction. Kinetic stability refers to reactivity or the ability of the metal complex to undergo ligand substitution reactions. Complexes which undergo extremely rapid ligand substitution reaction are referred to as labile complexes, and complexes that undergo extremely slow ligand substitution reaction are referred to as inert complexes. Sometimes the thermodynamic and kinetic stabilities of complexes are parallel to one another, but often they do not. One of the suitable examples for thermodynamically stable and kinetically inert complex is $[Ni(CN)_4]^{2-}$ as it undergoes ligand substitution reaction very rapidly. On the other hand, the cobalt complex $[Co(NH_3)_6]^{3+}$ is thermodynamically unstable but kinetically inert. The complex $[Co(NH_3)_6]^{3+}$ is thermodynamically unstable since the complex was observed to decompose very rapidly with rate in the order of 10^{25} in acidic solution. However, no ligand substitution reaction is found when the complex is kept in acidic solution for several days; hence the complex is kinetically inert. From the above two examples, it can be interpreted that the stability of a complex mainly depends upon the conditions, and it is always recommended to specify the conditions such as pH, temperature, etc. while mentioning the stability of the complex. In brief, it is not necessary for a stable complex to be inert and an unstable complex to be labile.

Consider the three complexes $[Ni(CN)_4]^{2-}$, $[Mn(CN)_6]^{3-}$, and $[Cr(CN)_6]^{3-}$. All the complexes are thermodynamically stable, but kinetically they behave in a different manner. The rate of exchange can be measured when carbon-14-labelled cyanide ions are reacted with metal complexes in solution. It indicates that $[Ni(CN)_4]^{2-}$ is labile, $[Mn(CN)_6]^{3-}$ is less labile, and $[Cr(CN)_6]^{3-}$ is inert and proves that not all stable complexes are inert and vice versa.

$$[Ni(CN)_4]^{2-} + 4\,{}^{14}CN^- \xrightarrow[\text{fast}]{\text{very}} [Ni({}^{14}CN)_4]^{2-} + 4CN^-$$

$$t_{1/2} = 30\text{sec}$$

$$[Mn(CN)_6]^{3-} + 6\,{}^{14}CN^- \xrightarrow{\text{slow}} [Mn({}^{14}CN)_6]^{3-} + 6CN^-$$

$$t_{1/2} = 1\text{hr}$$

$$[Cr(CN)_6]^{3-} + 6\,{}^{14}CN^- \xrightarrow{\text{slowest}} [Cr({}^{14}CN)_6]^{3-} + 6CN^-$$

$$t_{1/2} = 24\text{days}$$

3. Factors affecting the stability of metal complexes

There are several factors that can affect the stability of the metal complexes [2, 5, 8, 9], which include:

1. Nature of the central metal ion
2. Nature of the ligand
3. Chelating effect
4. Macrocyclic effect
5. Resonance effect
6. Steric effect or steric hindrance

3.1 Nature of central metal ion

3.1.1 Charge on metal cation

In metal cations, higher oxidation state forms more stable complex than lower oxidation states with ligands such as NH_3, H_2O, etc. Even few exceptions are there like CO, PMe_3, o-phenanthroline, bipyridyl, CN^-, which form more stable complex with lower oxidation state metals.

3.1.2 Size of central metal cation

The stability of metal complex increases with decrease in size of the metal cations. For M^{2+} ions, the general trend in stability for complexes is

$$Ba^{2+} < Sr^{2+} < Ca^{2+} < Mg^{2+} < Mn^{2+} < Fe^{2+} < Co^{2+} < Ni^{2+} < Cu^{2+} > Zn^{2+}$$

This trend in stability is known as Irving-Williams series.

This order of stability is also in good agreement with the charge to radius ratio concept because the radii decrease from Ba^{2+} to Cu^{2+} and then increased to Zn^{2+}. The order of size of dipositive ions is

$$Ba^{2+} > Sr^{2+} > Ca^{2+} > Mg^{2+} > Mn^{2+} > Fe^{2+} > Co^{2+} > Ni^{2+} > Cu^{2+} < Zn^{2+}$$

3.2 Nature of ligands

Basic character of ligands: The greater is the basic character of ligand, the more easily it can donate its lone pair of electrons to the central metal ion and hence greater is the complex stability. In 3D-series metal ion, order of stability of complex with NH_3, H_2O, and F^- is:

$$NH_3 > H_2O > F^-$$

The nature of metal-ligand bond also affects the stability of metal complexes. The higher the covalent character, the greater will be the complex stability. For example, the stabilities of silver complexes have different halide ligands which are in the following order:

$$AgI_2^- > AgBr_2^- > AgCl_2^- > AgF_2^-$$

Ligands having vacant p- or d-orbital tend to form π bond and hence form stable complexes with metals. Ligands that are capable of forming such π bond are CO, CN^-, alkene, phenanthroline, etc.

3.3 The chelate effect

The chelate effect is that the complexes resulting from coordination of metal ions with the chelating ligand are thermodynamically much more stable than the complexes with non-chelating ligands [10, 11]. Chelating ligands are molecules which can bind to single metal ion through several bonds and are also called as multidentate ligands. Simple (and common) examples include ethylenediamine and oxalate. Non-chelating ligands are ligands that bond to just one site, such as chloride, cyanide, and water.

The chelate effect can be understood by comparing the reaction of a metal ion, respectively, with a chelating ligand and with a monodentate ligand having similar/comparable donating groups. During the comparison study, the number of coordination should be maintained equal in both the cases, for example, the value obtained while adding a bidentate ligand is compared with the value obtained for two monodentate ligands. For example, coordination of metal ion with chelating ligand 2,2′-bipyridine can be compared with that of monodentate pyridine ligand. Another such comparison can be made between coordination behavior of chelating 1,2-diaminoethane (ethylenediamine = en) and monodentate ammonia. Such comparison studies revealed that the metal complex formed from chelating ligands are thermodynamically more stable than the complex formed from monodentate ligand. For example, formation of complexes from hydrated cadmium ion, $[Cd(H_2O)_4]^{2+}$ with methylamine (CH_3NH_2), ethylenediamine (en) and triethylenetetramine (trien), and their stability is in the following order:

$$[Cd(CH_3NH_2)_4]^{2+} < [Cd(en)]^{2+} < [Cd(trien)]^{2+}$$

3.4 Macrocyclic effect

A macrocyclic ligand is a cyclic molecule that contains nine or more atoms in the cyclic structure and has three or more potential donor atoms which can coordinate to the metal ion. It has been observed that the stability of metal complexes in the presence of macrocyclic ligand of appropriate size is higher than the stability of complexes coordinated to open-ended multidentate chelating ligands. Some notable examples of macrocyclic ligands include cyclic crown polyether, heme, etc. [12].

3.5 Resonance effect

Resonance increases the stability of the complexes. For example, acetylacetonate anion ligand shows resonance, and as a result it forms stable complexes upon reacting with metal ion (**Figure 1**). The ligand-metal π bonding increases the delocalization of electrons compared to free enolate as shown below and leads to increased stability (**Figure 2**).

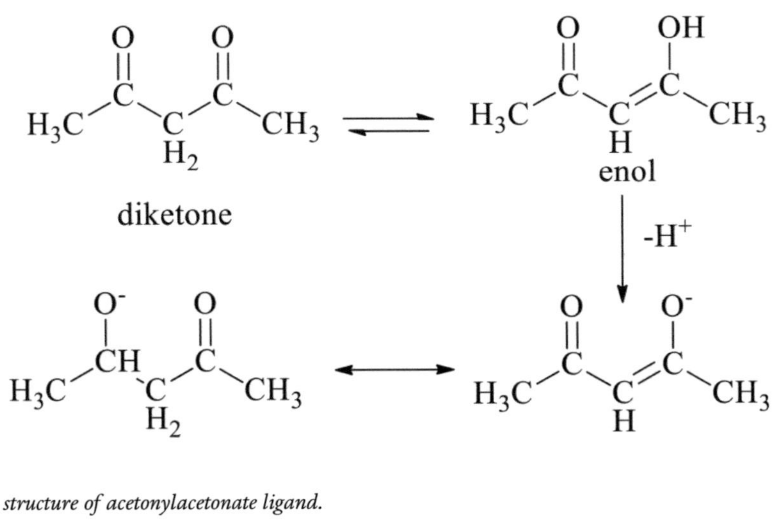

Figure 1.
Resonance structure of acetonylacetonate ligand.

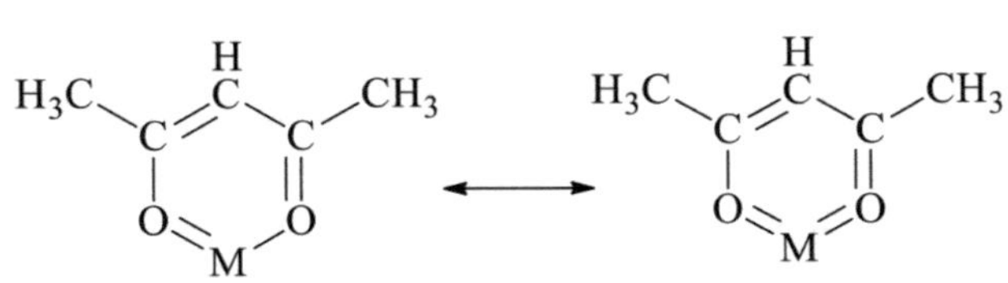

Figure 2.
Acetonylacetonate-metal complex.

3.6 Steric effect

The presence of bulky substituents in the ligands can affect the stability of the metal complex, and this type of destabilization of metal complex due to bulkiness of the substituent is called as steric effect [13]. For example, consider the ligand 8-hydroxy quinoline and its methyl substituted derivative 2-methyl-8-hydroxy quinolone. Both are bidentate ligands and form chelated complexes with Ni^{2+} ion as shown in **Figure 3**.

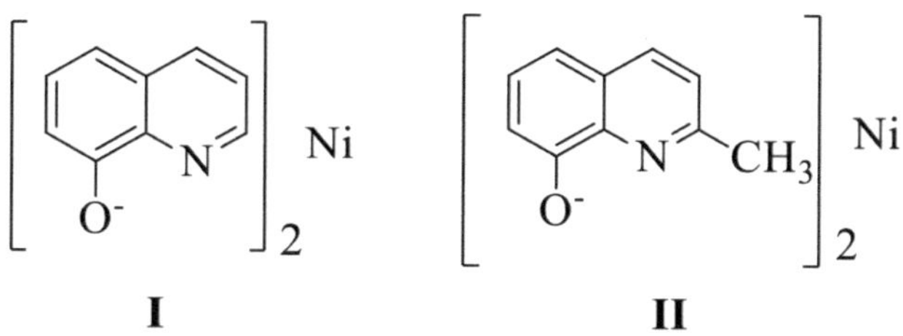

Figure 3.
Chelating complexes of Ni(II) ion showing steric effect.

The complex (II) is less stable than complex (I) because of bulky group attached to an atom adjacent to donor atom which cause a steric hindrance and lower the stability of the complex.

3.7 Crystal field stabilization energy (CFSE)

The crystal field stabilization energy (CFSE) is one of the most important factors that decides the stability of the metal complexes. CFSE is the stability that arises when a metal ion coordinates to a set of ligands, which is due to the generation of a crystal field by the ligands. Thus, a higher value of CFSE means that the complex is thermodynamically stable and kinetically inert. Some of the notable examples of complexes that have high CFSE are low spin $5d^6$ complexes of Pt^{4+} and Ir^{3+} and square planar $5d^8$ complexes of Pt^{2+}. All these complexes are thermodynamically stable and kinetically inert, which undergo ligand substitution reactions extremely slowly [3, 4].

4. Determination of stability constants

The determination of metal complexes involves several methods including spectroscopic and potentiometric methods. The determination of stability constant is very significant to understand the role and behavior of ligand(s) in stabilizing the metal complexes and found applications in the fields of biology, environmental study, metallurgy, food chemistry, and many other industrial processes. Some of the methods that are used for the determination of stability constants are given as follows.

4.1 Spectroscopic methods

UV-Vis spectroscopic technique has been used to determine the stability constant and composition of a complex [14]. The formation of metal complex is indicated by the change in absorbance in the UV-Vis spectroscopy. The relationship between absorbance (A) and concentration is given by Beer's law as shown.

$$A = \varepsilon.c.l.$$

where ε = molar extinction coefficient, l = path length of the absorption cell, c = concentration of the complex.

From the above equation, the concentration (c) of the metal complex can be calculated by measuring absorbance (A) using spectrophotometer and knowing the molar extinction coefficient (ε) at that wavelength (λ) and path length (l). For example, consider the formation constant (K_f) for the following reaction:

$$M^{n+} + L \xrightarrow{K_f} ML^{n+}, \quad K_f = \frac{[ML^{n+}]}{[M^{n+}][L]} \tag{7}$$

By knowing the values of $[ML^{n+}]$, [M], and [L], Kf can be calculated. $[ML^{n+}]$, [M], and [L] can be calculated as follows:

$$C_M = [M^{n+}] + [ML^{n+}] \tag{8}$$

$$C_L = [L] + [ML^{n+}] \tag{9}$$

where C_L and C_M are the total concentrations of the ligand and metal ion, respectively.

From Beer's law

$$A = \sum_{ML^{n+}} \times l \times [ML^{n+}] \tag{10}$$

On rearranging

$$[ML^{n+}] = A/\sum_{ML^{n+}} \times l \tag{11}$$

On substituting Eq. (11) in (8), we get

$$C_M = [M^{n+}] + A/\sum_{ML^{n+}} \times l$$

$$\text{or } [M^{n+}] = C_M - A/\sum_{ML^{n+}} \times l \tag{12}$$

Similarly, from Eqs. (9) and (11), we get

$$[L] = C_L - A/\sum_{ML^{n+}} \times l \tag{13}$$

By introducing the values of $[ML^{n+}]$, $[M^{n+}]$, and [L] from Eqs. (11) to (13) in the formation constant Eq. (7), we can determine the value of K_f.

4.2 Job's method or method of continuous variations (MCV)

Job's method of continuous variations (MCV) is used to determine the complex formation as well as stability constants [14, 15]. Job's method is basically used to determine the composition of metal complexes, and this is the modified version of spectroscopic method. This method is applicable in the case of solutions, where the formation of one metal complex takes place.

This method includes the following steps:

i. Make a volume of 10 mL solutions of metal complex containing different proportions of metal ion as well as ligand. The number of solutions should be 10.

Sr. no. of solution:	1	2	3	4	5	6	7	8	9	10
Volume of metal ion (mL)	0	1	2	3	4	5	6	7	8	9
Volume of ligand (mL)	9	8	7	6	5	4	3	2	1	0

Here the total concentration of the solution is constant, i.e., sum of concentration of the metal, C_M, and the ligand, C_L, is fixed. Therefore:

$$C_M + C_L = C \text{ (constant)} \tag{14}$$

ii. In the next step, optical densities of the solutions prepared in the previous step are to be determined spectrophotometrically at the wavelength of light which is strongly absorbed by metal complex but does not get absorbed by metal ion and ligand.

iii. A plot between mole fraction (*mf*) of the ligand and absorbance or optical density is to be drawn. The plot obtained is shown in **Figure 4**.

On the extrapolation of the curve, the legs of the curve intersect each other at a point which is the point of maximum absorbance.

Suppose ML_n is the formula of the complex, then

$$n = C_L/C_M \tag{15}$$

Eq. (14) can be rewritten as:

$$C_M/C + C_L/C = C/C = 1 \tag{16}$$

From this equation, namely,

$$C_L/C = mf \tag{17}$$

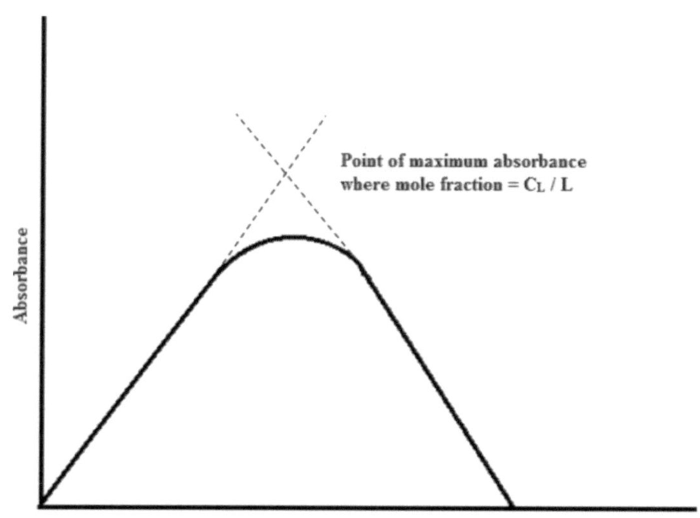

Figure 4.
Determination of the composition of metal complex Job's method of continuous variation.

On reducing Eq. (16), we get.

$$mf + C_M/C = 1$$

or $$C_M/C = 1 - mf \quad (18)$$

By dividing (17) by (18), we get.

$$C_L/C \text{ X } C/C_M = mf/1 - mf$$

or $$C_L/C_M = mf/1 - mf$$

or $$n = mf/1 - mf \quad (19)$$

The composition of the complex, ML_n, can be determined by the value of n as obtained from Eq. (19). There are some drawbacks of this method. One of the drawbacks is that this method is applicable where, under experimental conditions, only one complex is formed. Also, the total volume of the solutions which contain metal ion and ligand should not be changed. The method of continuous variations has also found tremendous application in the field of organometallic chemistry [16].

4.3 Bjerrum's method

The type and extent of interaction existing between the metal ion and ligand can be investigated by various experimental methods [17–21], and each method requires different experimental conditions and resulted in differences in the interpretation of reaction mechanism and stability constants. Bjerrum's method to determine the stability constant is also known as potentiometric method. This method is based on the competition between hydrogen ion and metal ion for ligand which is a weak base. Consider a metal ion and an acid such as nitric acid are added to a ligand in aqueous solution; the following equations are obtained:

When acid reacts with ligand

$$L + H^+ \xrightleftharpoons{K_a} HL^+; \; K_a = \frac{[HL^+]}{[L][H^+]}$$

where K_a is acid association constant of the ligand

When metal ion reacts with ligand

$$L + M^+ \xrightleftharpoons{K_f} ML^+; \; \text{where } K_f = \frac{[ML^+]}{[L][M^+]} \quad (20)$$

where K_f is the formation constant.

Let us consider C_H, C_M, and C_L are the total concentrations of acid, metal ion, and ligand, respectively. Then

$$C_H = [H^+] + [HL^+]$$

$$C_L = [L] + [ML^+] + [HL^+]$$

$$C_M = [M^+] + [ML^+]$$

By solving above three equations and using the acid association constant K_a, we get

$$[ML^+] = C_L - C_H + [H^+] - \frac{C_H - [H^+]}{K_a[H^+]}$$

$$[M^+] = C_M - [ML^+]$$

$$[L] = \frac{C_H - [H^+]}{K_a[H^+]}$$

The concentration terms of $[ML^+]$, $[M^+]$, and [L] can be calculated potentiometrically by using a pH meter and by substituting the values of $[ML^+]$, $[M^+]$, and [L] in the formation constant (K_f); Eq. (20) can determine the value of K_f.

In the potentiometric or pH-metric determination of stability constant, a ligand and an acid such as nitric acid are titrated against standard NaOH during which period the pH of the solution has to be measured after each addition. Throughout the experimental studies, several conditions must be maintained such as

For electrical neutrality

$$[H^+] + [Na^+] = [OH^-] + [L^-] \tag{21}$$

The total concentration of the ligand is given as follows:

$$L_{Total} = [HL] + [L^-] \tag{22}$$

The protonation of the ligand can be represented as

$$HL \rightleftharpoons H^+ + L^-$$

The equilibrium constant K_a is given as

$$K_a = \frac{[H^+][L^-]}{[HL]} \tag{23}$$

By combining Eqs. (21)–(23), we get

$$pK_a = -\log[H^+] + \log\left(\frac{L_{Total} - \{[Na^+] + [H^+] - [OH^-]\}}{[Na^+] + [H^+] - [OH^-]}\right)$$

$[OH^-]$ and $[H^+]$ are negligible as compared to $[Na^+]$ at near neutral solution. Hence Eq. (24) becomes

$$pK_a = -\log[H^+] + \log\left(\frac{L_{Total} - [Na^+]}{[Na^+]}\right)$$

The above equation is used to calculate the value of pK_a of ligand.

When a ligand is titrated with a solution containing M^{n+} ion and an equivalent amount of H^+, the resulting solution will have an equilibrium mixture of H^+, OH^-, HL, L^-, M^{n+}, $M^{(n-1)+}L$, $M^{(n-2)+}L_2$, etc. By measuring the pH values with the help of pH meter and knowing the concentrations of M^{n+}, H^+, HL, etc., one can calculate the stepwise stability constants K_1, K_2, K_3, etc. During the evaluation of equilibrium constants, the concentrations of M^{n+} and L^- are varied, and such variations in the concentration will lead to changes in the ionic strength of the solutions. In order to maintain the constant ionic strength, a large excess of an ionic salt is added to the reaction mixture. The presence of large excess of ionic salt will compensate any changes in the ionic strength of the solution. The ionic salts that are added for such purpose should not react with M^{2+} or L^-, and commonly used salts include KNO_3 and $NaClO_4$, due the low affinity of NO_3^- and ClO_4^- ions for most of the M^{2+} ions. For example, KNO_3 was added in excess during the binding study of the ligand p-aminobenzoic acid with Ni, Mg, and Co metal ions. The p-aminobenzoic acid has two coordination sites such as amino and carboxylate groups and has a pK_a value of 5.9153. The stability constant values obtained for Ni, Mg, and Co complexes are depicted in **Table 1**.

The stability constant values for Ni shows the trend 2:3 < 1:5 < 1:1 < 1:2, while the trend for Co is 1:2 < 2:3 < 1:1 < 1:5 and for Mg it is 1:5 < 1:2 < 1:1 < 2:3. The values obtained from the above study indicates that 1:2 complex of Ni complex is more stable, whereas Co complex is stable in 1:5 ratio and that of Mg is more stable in the ratio of 2:3.

4.4 Irving and Rossotti method

The Irving and Rossotti method for the determination of stability constant is also based on the principle of potentiometric method [21]. Using this method, the formation curve of metal complex can directly be calculated with the help of pH meter. Another major advantage of this method over the Bjerrum's method is that the calculation is simple and does not require hydrogen ion concentration. Moreover, this method can be used for types of ligands that are conjugate to weak acids. The calculation of stability constant using this method involves the following steps.

Step 1: The following solutions were titrated separately against base solution

a. Titration with free acid (A)

b. Titration with free acid + ligand (A + L)

c. Titration with free acid + ligand + metal (A + L + M)

Step 2: Calculation of formation functions n, nA, and PL using the values used/obtained from above three titrations

The term formation function "n," also called as ligand number, is defined as the average number of ligands attached per metal center and is calculated using the following equation

$$n = \frac{\text{Total number of ligands coordinated to metal}}{\text{Total number of metal}}$$

Metal	M:L ratio			
	1:1	1:2	1:5	2:3
Ni	8.492	14.8593	8.3598	3.4649
Mg	8.4664	8.3392	7.0794	11.1943
Co	8.590	5.3186	8.6337	6.2330

Table 1.
Stability constant values obtained for Ni, Mg and Co complexes having different metal to ligand ratio.

The term nA is similar to n and is defined as the average number of protons bound to the ligand which are not coordinated to the metal center. PL gives the free ligand exponent. All the three terms n, nA, and PL can be calculated with the help of following equations

$$nA = \gamma - \frac{(V_2 - V_1)(N + \mathcal{E}^0)}{(V_0 + V_1)T^0L} \tag{25}$$

$$n = \frac{(V_3 - V_2)(N + \mathcal{E}^0)}{(V_0 + V_2)nAT^0M} \tag{26}$$

$$PL = \log\left\{1 + \frac{\frac{[H^+]}{K_2} + \frac{[H^+]^2}{K_1K_2}}{(T^0L - T^0M)\overline{n}} X \frac{(V_0 + V_3)}{V_0}\right\} \tag{27}$$

where N is the normality of base used; V_0 is the initial volume of the solution; V_1, V_2, and V_3 are the volume of base consumed during the (A), (A + L), and (A + L + M) titrations, respectively, at same pH value; T^0L is the initial concentration of ligand; $\mathcal{E}^0$ is the initial concentration of acid; γ is the number of titrable or replaceable protons.

Step 3: Determination of formation curves: by plotting formation function (n) against PL and nA against pH for a HL (protonated ligand) system.

The value formation constants corresponding to formation of protonated ligand are obtained by plotting nA against pH. Similarly, the stepwise stability constants for the formation of metal complexes are obtained from the formation curve resulted by plotting n against PL.

5. Conclusions

The thermodynamic and kinetic stability of coordination compounds along with the various factors affecting the stability of metal complexes have been discussed in this chapter. Stability constant and its determination have also been listed.

Acknowledgements

Authors acknowledge National Institute of Technology Kurukshetra, Haryana, India, for its support.

Conflict of interest

There is no conflict of interest.

Author details

Senthilkumar Muthaiah*, Anita Bhatia and Muthukumar Kannan
National Institute of Technology Kurukshetra, Kurukshetra, Haryana, India

*Address all correspondence to: msenthil@nitkkr.ac.in

References

[1] Kettle SFA. Chapter 5: Stability of Coordination Compounds: Physical Inorganic Chemistry. Berlin, Heidelberg: Springer-Verlag; 1996. p. 73

[2] Irving H, Williams RJP. The stability of transition metal complexes. Journal of the Chemical Society. 1953:3192

[3] Burgot J-L. Stability of complexes: Some elements concerning the kinetics of their formation. In: Ionic Equilibria in Analytical Chemistry. New York, NY: Springer; 2012. p. 439

[4] Huheey JE, Keiter EL, Keiter RL. Inorganic Chemistry: Principles of Structure and Reactivity. 4th ed. New York, NY: HarperCollins College Publishers; 1993

[5] Malik WH, Tuli GD, Madan RD. Selected Topics in Inorganic Chemistry. New Delhi: S. Chand & Co Ltd.; 2009

[6] Martell E, Hancock RD. Stability constants and their measurement. In: Metal Complexes in Aqueous Solutions. Modern Inorganic Chemistry. Boston, MA: Springer; 1996. p. 217

[7] Eigen M, Wilkins RG. Chapter 3: The kinetics and mechanisms of formation of metal complexes. In: Mechanisms of Inorganic Reactions. Vol. 49. American Chemical Society; 1965. p. 55

[8] Hancock RD, Nakani BS. Some factors influencing the stability of complexes with ligands containing neutral oxygen donor ligands, including Crown Ethers. Journal of Coordination Chemistry. 1984;**13**:309

[9] Martell AE, Hancock RD, Motekaitis RJ. Factors affecting stabilities of chelate, macrocyclic and macrobicyclic complexes in solution. Coordination Chemistry Reviews. 1994;**133**:39

[10] Frausto da Silva JJR. The chelate effect redefined. Journal of Chemical Education. 1983;**60**:390

[11] Munro D. Misunderstandings over the chelate effect. Chemistry in Britain. 1977;**13**:100

[12] Hendrick K, Tasker PA, Linday CF. The specification of bonding cavities in macrocylic ligands. Progress in Inorganic Chemistry. 1985;**33**:1

[13] Brown TL, Lee KJ. Ligand steric properties. Coordination Chemistry Reviews. 1993;**128**:89

[14] Job P. Formation and stability of inorganic complexes in solution. Annali Di Chimica Applicata. 1928;**9**:113

[15] Irving H, Pierce TB. Observations on Job's method of continuous variations and it's extension to two-phase system. Journal of the Chemical Society. 1959:2565

[16] Renny JS, Tomasevich LL, Tallmadge EH, Collum DB. Method of continuous variations: Applications of Job plots to the study of molecular associations in organometallic chemistry. Angewandte Chemie (International Edition in English). 2013;**52**(46):11998

[17] Martell AE, Motekaitis RJ. Potentiometry revisited: The determination of thermodynamic equilibria in complex multicomponent systems. Coordination Chemistry Reviews. 1990;**100**:323

[18] Al-Rashdi AA, Naggar AH, Farghaly OA, Mauof HA, Ekshiba AA. Potentiometric determination of stability constants of sulphathiazole and glycine-metal complexes. American Journal of Analytical Chemistry. 2018;**9**:99-112

[19] Ansari FB, Mansoora F, Quadria SH. Potentiometric determination of stability constants of transition metal complexes with P-aminobenzoic acid.

International Journal of Chemical Sciences. 2008;**6**(2):873

[20] Mohod SO, Tayade DT. pH metric study of phenylthiocarbamidophenol with Cu(II), Co(II), Cd(II) and Ni(II) metal ions at 0.1 M ionic strength. International Journal of Pharmaceutical Research. 2017;**8**(3):25

[21] Irving H, Williams RJP. Order of stability of metal complexes. Nature. 1948;**162**:746

Chapter 3

Stability Constants of Metal Complexes in Solution

Jagvir Singh, Abhay Nanda Srivastav, Netrapal Singh and Anuradha Singh

Abstract

In the formation of metal complexes in an aqueous medium, equilibrium constant or stability constant is used to determine the strength of interaction between reagents that make the final product after the formation of bonds. In general stability means that a complex may be stored for a long time under suitable conditions or this compound may be existing under suitable conditions. Regarding how much is the concentration of complexes in solution, stability constant provides this information via calculations. These calculations are very much important in many areas of science like chemistry, biology, and medicine. During the complex formation in aqueous medium, two types of stabilities are considered: one is the thermodynamic stability, and the other is kinetic stability. Stability of metal complexes may be affected by various factors like nature of central metal ion and ligand, chelating effect, etc., and some parameters like distribution coefficients, conductance, refractive index, etc. are useful for the determination of stability constants. Various modern techniques are used to determine the stability constant of simple as well as mixed ligand compounds.

Keywords: thermodynamic stability, kinetic stability, chelate effect, distribution method, ion exchange method, Bjerrum's method

1. Introduction

Stability constant of the formation of metal complexes is used to measure interaction strength of reagents. From this process, metal ion and ligand interaction formed the two types of metal complexes; one is supramolecular complexes known as host-guest complexes [1] and the other is anion-containing complexes. In the solution it provides and calculates the required information about the concentration of metal complexes.

Solubility, light, absorption conductance, partitioning behavior, conductance, and chemical reactivity are the complex characteristics which are different from their components. It is determined by various numerical and graphical methods which calculate the equilibrium constants. This is based on or related to a quantity, and this is called the complex formation function.

During the displacement process at the time of metal complex formation, some ions disappear and form a bonding between metal ions and ligands. It may be considered due to displacement of a proton from a ligand species or ions or molecules causing a drop in the pH values of the solution [2]. Irving and Rossotti developed a technique for the calculation of stability constant, and it is called potentiometric technique.

To determine the stability constant, Bjerrum has used a very simple method, and that is metal salt solubility method. For the studies of a larger different variety of polycarboxylic acid-, oxime-, phenol-containing metal complexes, Martel and Calvin used the potentiometric technique for calculating the stability constant. Those ligands [3, 4] which are uncharged are also examined, and their stability constant calculations are determined by the limitations inherent in the ligand solubility method. The limitations of the metal salt solubility method and the result of solubility methods are compared with this. M-L, MLM, and (M3) L are some types of examples of metal-ligand bonding. One thing is common, and that is these entire types metal complexes all have one ligand.

The solubility method can only usefully be applied to studies of such complexes, and it is best applied for ML; in such types of system, only ML is formed. Jacqueline Gonzalez and his co-worker propose to explore the coordination chemistry of calcium complexes. Jacqueline and et al. followed this technique for evaluate the as partial model of the manganese-calcium cluster and spectrophotometric studies of metal complexes, i.e., they were carried calcium(II)-1,4-butanediamine in acetonitrile and calcium(II)-1,2-ethylendiamine, calcium(II)-1,3-propanediamine by them.

Spectrophotometric programming of HypSpec and received data allows the determination of the formation of solubility constants. The logarithmic values, log β_{110} = 5.25 for calcium(II)-1,3-propanediamine, log β_{110} = 4.072 for calcium(II)-1,4-butanediamine, and log β_{110} = 4.69 for calcium(II)-1,2-ethylendiamine, are obtained for the formation constants [5]. The structure of Cimetidine and histamine H2-receptor is a chelating agent. Syed Ahmad Tirmizi has examined Ni(II) cimetidine complex spectrophotometrically and found an absorption peak maximum of 622 nm with respect to different temperatures.

Syed Ahmad Tirmizi have been used to taken 1:2 ratio of metal and cimetidine compound for the formation of metal complex and this satisfied by molar ratio data. The data, $1.40–2.4 \times 10^8$, was calculated using the continuous variation method and stability constant at room temperature, and by using the mole ratio method, this value at 40°C was $1.24–2.4 \times 10^8$. In the formation of lead(II) metal complexes with 1-(aminomethyl) cyclohexene, Thanavelan et al. found the formation of their binary and ternary complexes. Glycine, L-proline, L-alanine, L-isoleucine, L-valine, and L-leucine are α-amino acids, and these are important biologically [6]. These α-amino acids are also investigated by potentiometric technique at 32°C. The mixed ligands were also studied using these methods. 50% (v/v) DMSO-water medium used for the determination of acidity constants and their stability constants these type ligands. In a stepwise manner, the ternary complexes were synthesized.

Using the stability constant method, these ternary complexes were found out, and using the parameters such as $\Delta \log K$ and $\log X$, these ternary complex data were compared with binary complex. The potentiometric technique at room temperature (25°C) was used in the investigation of some binary complex formations by Abdelatty Mohamed Radalla. These binary complexes are formed with 3D transition metal ions like Cu^{2+}, Ni^{2+}, Co^{2+}, and Zn^{2+} and gallic acid's importance as a ligand and 0.10 mol dm^{-3} of $NaNO_3$. Such types of aliphatic dicarboxylic acids are very important biologically. Many acid-base characters and the nature of using metal complexes have been investigated and discussed time to time by researchers [7].

The above acids (gallic and aliphatic dicarboxylic acid) were taken to determine the acidity constants. For the purpose of determining the stability constant, binary and ternary complexes were carried in the aqueous medium using the experimental conditions as stated above. The potentiometric pH-metric titration curves are inferred for the binary complexes and ternary complexes at different ratios, and

DOI: http://dx.doi.org/10.5772/intechopen.90183

formation of ternary metal complex formation was in a stepwise manner that provided an easy way to calculate stability constants for the formation of metal complexes.

The values of Δ log *K*, percentage of relative stabilization (% R. S.), and log *X* were evaluated and discussed. Now it provides the outline about the various complex species for the formation of different solvents, and using the concentration distribution, these complexes were evaluated and discussed. The conductivity measurements have ascertained for the mode of ternary chelating complexes.

A study by Kathrina and Pekar suggests that pH plays an important role in the formation of metal complexes. When epigallocatechin gallate and gallic acid combine with copper(II) to form metal complexes, the pH changes its speculation. We have been able to determine its pH in frozen and fluid state with the help of multifrequency EPR spectroscopy [8]. With the help of this spectroscopy, it is able to detect that each polyphenol exhibits the formation of three different mononuclear species. If the pH ranges 4–8 for di- or polymeric complex of Cu(II), then it conjectures such metal complexes. It is only at alkaline pH values.

The line width in fluid solutions by molecular motion exhibits an incomplete average of the parameters of anisotropy spin Hamilton. If the complexes are different, then their rotational correlation times for this also vary. The analysis of the LyCEP anisotropy of the fluid solution spectra is performed using the parameters determined by the simulation of the rigid boundary spectra. Its result suggests that pH increases its value by affecting its molecular mass. It is a polyphenol ligand complex with copper, showing the coordination of an increasing number of its molecules or increasing participation of polyphenol dimers used as ligands in the copper coordination region.

The study by Vishenkova and his co-worker [8] provides the investigation of electrochemical properties of triphenylmethane dyes using a voltammetric method with constant-current potential sweep. Malachite green (MG) and basic fuchsin (BF) have been chosen as representatives of the triphenylmethane dyes [9]. The electrochemical behavior of MG and BF on the surface of a mercury film electrode depending on pH, the nature of background electrolyte, and scan rate of potential sweep has been investigated.

Using a voltammetric method with a constant-current potential sweep examines the electrical properties of triphenylmethane dye. In order to find out the solution of MG and BF, certain registration conditions have been prescribed for it, which have proved to be quite useful. The reduction peak for the currents of MG and BF has demonstrated that it increases linearly with respect to their concentration as 9.0×10^{-5}–7.0×10^{-3} mol/dm^3 for MG and 6.0×10^{-5}–8.0×10^{-3} mol/dm^3 for BF and correlation coefficients of these values are 0.9987 for MG and 0.9961 for BF [10].

5.0×10^{-5} and 2.0×10^{-5} mol/dm^3 are the values used as the detection limit of MG and BF, respectively. Stability constants are a very useful technique whose size is huge. Due to its usefulness, it has acquired an umbrella right in the fields of chemistry, biology, and medicine. No science subject is untouched by this. Stability constants of metal complexes are widely used in the various areas like pharmaceuticals as well as biological processes, separation techniques, analytical processes, etc. In the presented chapter, we have tried to explain this in detail by focusing our attention on the applications and solutions of stability of metal complexes in solution.

2. Stability constant of metal complexes

Stability or formation or binding constant is the type of equilibrium constant used for the formation of metal complexes in the solution. Acutely, stability constant is applicable to measure the strength of interactions between the ligands and metal ions that are involved in complex formation in the solution [11]. A generally these 1-4 equations are expressed as the following ways:

$$\text{Metal} + \text{Ligand} \leftrightarrows \text{Metal} - \text{Ligand}\ K_1 = \frac{(ML)}{[M][L]} \tag{1}$$

$$\text{Metal} + \text{Ligands} \leftrightarrows \text{Metal} + \text{Ligand}_2\ K_2 = \frac{(ML_2)}{[ML][L]} \tag{2}$$

$$\text{Metal} + \text{Ligand}_3 \leftrightarrows \text{Metal} + \text{Ligand}_3\ K_3 = \frac{(ML_3)}{[ML_2][L]} \tag{3}$$

Thus

$$\text{Metal} + \text{Ligand}_{n-1} + L \leftrightarrows \text{Metal} + \text{Ligand}_{-n}\ K_n = \frac{(ML_n)}{[ML_{n-1}][L]} \tag{4}$$

K_1, K_2, K_3, ... K_n are the equilibrium constants and these are also called stepwise stability constants. The formation of the metal-ligand$_{-n}$ complex may also be expressed as equilibrium constants by the following steps:

$$\text{Metal} + \text{Ligand} \xrightarrow{B_1} \text{Metal} - \text{Ligand}, \beta = \frac{(ML)}{[M][L]} \tag{5}$$

$$\text{Metal} + 2\text{Ligand} \xrightarrow{B_2} \text{Metal} - \text{Ligand}_2, \beta_2 = \frac{(ML_2)}{[M][L]^2} \tag{6}$$

$$\text{Thus Metal} + n\text{Ligand} \xrightarrow{B_n} \text{Metal} - \text{ligandLn}, \beta_n = \frac{(MLn)}{[M][L]^n} \tag{7}$$

β_1, β_2, β_3, ... β_n are the equilibrium constants, and these equilibrium constants are known as overall stability constants or overall formation. β_n is called as the nth cumulative or overall formation constant [12]. Any metal complexes will be of greater stability if its stability constant has the higher value. Sometimes the 1/k values are alternative values of stability constant, and now this is called as instability constant. $Log_{10}K_1$, $log_{10}K_2$ ... $log_{10}K_n$, and $log_{10}\beta_n$ are the ways that expressed the stepwise and cumulative stability constants.

3. Relationship or interaction between β_n and K_1, K_2, K_3, ... K_n

The parameters K and β are related together, and these are expressed in the following example:

$$\beta_3 = \frac{(ML_3)}{[M][L]3} \tag{8}$$

Now the numerator and denominator are multiplied together with the use of [metal-ligand] [metal-ligand$_2$], and after the rearranging we get the following equation:

$$\begin{aligned}\beta_3 &= \frac{[ML_3]}{[M][L]^3} \times \frac{[ML][ML_2]}{[ML][ML_2]} \\ &= \frac{[ML_1]}{[M][L]} \times \frac{[ML_2]}{[ML][L]} \times \frac{[ML_3]}{[ML_2][L]} = K_1 \text{ x } K_2 \text{ x } K_3\end{aligned} \tag{9}$$

$$\text{Thus } \beta_n = \frac{[ML_1]}{[M][L]} \times \frac{[ML_2]}{[ML][L]} \cdots \frac{[ML_n]}{[ML_{n-1}][L]} = K_1 \text{ x } K_2 \ldots K_n \tag{10}$$

Now we expressed it as the following:

$$\beta_n = \sum_{n=1}^{n=n} K_n \tag{11}$$

From the above relation, it is clear that the overall stability constant β_n is equal to the product of the successive (i.e., stepwise) stability constants, K_1, K_2, K_3, … K_n. This in other words means that the value of stability constants for a given complex is actually made up of a number of stepwise stability constants. The term stability is used without qualification to mean that the complex exists under a suitable condition and that it is possible to store the complex for an appreciable amount of time. The term stability is commonly used because coordination compounds are stable in one reagent but dissociate or dissolve in the presence of another regent. It is also possible that the term stability can be referred as an action of heat or light or compound. The stability of complex [13] is expressed qualitatively in terms of thermodynamic stability and kinetic stability.

3.1 Thermodynamic stability

In a chemical reaction, chemical equilibrium is a state in which the concentration of reactants and products does not change over time. Often this condition occurs when the speed of forward reaction becomes the same as the speed of reverse reaction. It is worth noting that the velocities of the forward and backward reaction are not zero at this stage but are equal.

If hydrogen and iodine are kept together in molecular proportions in a closed process vessel at high temperature (500°C), the following action begins:

$$H_2 + I_2 \rightarrow 2HI \tag{12}$$

In this activity, hydrogen iodide is formed by combining hydrogen and iodine, and the amount of hydrogen iodide increases with time. In contrast to this action, if the pure hydrogen iodide gas is heated to 500°C in the reaction, the compound is dissolved by reverse action, which causes hydrogen iodide to dissolve into hydrogen and iodine, and the ratio of these products increases over time. This is expressed in the following reaction:

$$2HI \rightarrow H_2 + I_2 \tag{13}$$

For the formation of metal chelates, the thermodynamic technique provides a very significant information. Thermodynamics is a very useful technique in distinguishing between enthalpic effects and entropic effects. The bond strengths are totally effected by enthalpic effect, and this does not make any difference in the whole solution in order/disorder. Based on thermodynamics the chelate effect below can be best explained. The change of standard Gibbs free energy for equilibrium constant is response:

$$\Delta G = -2.303\ RT\ \log_{10} \beta. \tag{14}$$

Where:
R = gas constant
T = absolute temperature
At 25°C,

$\Delta G = (-5.708 \text{ kJ mol}^{-1}) \cdot \log \beta$.

The enthalpy term creates free energy, i.e.,

$$\Delta G = \Delta H - T\Delta S \quad (15)$$

For metal complexes, thermodynamic stability and kinetic stability are two interpretations of the stability constant in the solution. If reaction moves from reactants to products, it refers to a change in its energy as shown in the above equation. But for the reactivity, kinetic stability is responsible for this system, and this refers to ligand species [14].

Stable and unstable are thermodynamic terms, while labile and inert are kinetic terms. As a rule of thumb, those complexes which react completely within about 1 minute at 25°C are considered labile, and those complexes which take longer time than this to react are considered inert. $[Ni(CN)_4]^{2-}$ is thermodynamically stable but kinetically inert because it rapidly exchanges ligands.

The metal complexes $[Co(NH3)_6]^{3+}$ and such types of other complexes are kinetically inert, but these are thermodynamically unstable. We may expect the complex to decompose in the presence of acid immediately because the complex is thermodynamically unstable. The rate is of the order of 10^{25} for the decomposition in acidic solution. Hence, it is thermodynamically unstable. However, nothing happens to the complex when it is kept in acidic solution for several days. While considering the stability of a complex, always the condition must be specified. Under what condition, the complex which is stable or unstable must be specified such as acidic and also basic condition, temperature, reactant, etc.

A complex may be stable with respect to a particular condition but with respect to another. In brief, a stable complex need not be inert and similarly, and an unstable complex need not be labile. It is the measure of extent of formation or transformation of complex under a given set of conditions at equilibrium [15].

Thermodynamic stability has an important role in determining the bond strength between metal ligands. Some complexes are stable, but as soon as they are introduced into aqueous solution, it is seen that these complexes have an effect on stability and fall apart. For an example, we take the $[Co\ (SCN)_4]^{2+}$ complex. The ion bond of this complex is very weak and breaks down quickly to form other compounds. But when $[Fe(CN)_6]^{3-}$ is dissolved in water, it does not test Fe^{3+} by any sensitive reagent, which shows that this complex is more stable in aqueous solution. So it is indicated that thermodynamic stability deals with metal-ligand bond energy, stability constant, and other thermodynamic parameters.

This example also suggests that thermodynamic stability refers to the stability and instability of complexes. The measurement of the extent to which one type of species is converted to another species can be determined by thermodynamic stability until equilibrium is achieved. For example, tetracyanonickelate is a thermodynamically stable and kinetic labile complex. But the example of hexa-amine cobalt(III) cation is just the opposite:

$$\left[Co(NH_3)_6\right]^{3+} + 6H_3O^+ \rightarrow \left[Co(H_2O)_6\right]^{3+} + 6NH_4^+ \quad (16)$$

Thermodynamics is used to express the difference between stability and inertia. For the stable complex, large positive free energies have been obtained from ΔG_0 reaction. The ΔH_0, standard enthalpy change for this reaction, is related to the equilibrium constant, β_n, by the well thermodynamic equation:

$$\Delta G_0 = -RT\ \ln\ \beta \quad (17)$$

$$\Delta G_0 = \Delta H_0 - T\Delta S_0 \quad (18)$$

DOI: http://dx.doi.org/10.5772/intechopen.90183

For similar complexes of various ions of the same charge of a particular transition series and particular ligand, ΔS_0 values would not differ substantially, and hence a change in ΔH_0 value would be related to change in β_n values. So the order of values of ΔH_0 is also the order of the β_n value.

3.2 Kinetic stability

Kinetic stability is referred to the rate of reaction between the metal ions and ligand proceeds at equilibrium or used for the formation of metal complexes. To take a decision for kinetic stability of any complexes, time is a factor which plays an important role for this. It deals between the rate of reaction and what is the mechanism of this metal complex reaction.

As we discuss above in thermodynamic stability, kinetic stability is referred for the complexes at which complex is inert or labile. The term "inert" was used by Tube for the thermally stable complex and for reactive complexes the term 'labile' used [16]. The naturally occurring chlorophyll is the example of polydentate ligand. This complex is extremely inert due to exchange of Mg^{2+} ion in the aqueous media.

4. Factors affecting the stability of complexes

The nature of central atom of metal complexes, dimension, its degree of oxidation, electronic structure of these complexes, and so many other properties of complexes are affected by the stability constant. Some of the following factors described are as follows.

4.1 Nature of central metal ion

In the coordination chemistry, metal complexes are formed by the interaction between metal ions and ligands. For these type of compounds, metal ions are the coordination center, and the ligand or complexing agents are oriented surrounding it. These metal ions mostly are the transition elements. For the determination of stability constant, some important characteristics of these metal complexes may be as given below.

4.2 Ionic size

Ligands are oriented around the central metal ions in the metal complexes. The sizes of these metal ions determine the number of ligand species that will be attached or ordinated (dative covalent) in the bond formation. If the sizes of these metal ions are increased, the stability of coordination compound defiantly decreased. Zn(II) metal ions are the central atoms in their complexes, and due to their lower size (0.74A°) as compared to Cd(II) size (0.97A°), metal ions are formed more stable.

Hence, Al^{3+} ion has the greatest nuclear charge, but its size is the smallest, and the ion N^{3-} has the smallest nuclear charge, and its size is the largest [17]. Inert atoms like neon do not participate in the formation of the covalent or ionic compound, and these atoms are not included in isoelectronic series; hence, it is not easy to measure the radius of this type of atoms.

4.3 Ionic charge

The properties of stability depend on the size of the metal ion used in the complexes and the total charge thereon. If the size of these metal ions is small and the total charge is high, then their complexes will be more stable. That is, their ratio will depend on the charge/radius. This can be demonstrated through the following reaction:

$$Fe^{3+} + 6CN^- \leftrightarrows [Fe(CN)_6]^{3-} \ \log\beta = 31 \text{ (More Stable)} \quad (19)$$

$$Fe^{2+} + 6CN^- \leftrightarrows [Fe(CN)_6]^{4-} \ \log\beta = 8.3 \text{ (Less Stable)} \quad (20)$$

An ionic charge is the electric charge of an ion which is formed by the gain (negative charge) or loss (positive charge) of one or more electrons from an atom or group of atoms. If we talk about the stability of the coordination compounds, we find that the total charge of their central metal ions affects their stability, so when we change their charge, their stability in a range of constant can be determined by propagating of error [18]. If the charge of the central metal ion is high and the size is small, the stability of the compound is high:

$$Li^+ > Na^+ > K^+ > Rb^+ > Cs^+ \quad (21)$$

$$Th^{4+} > Y^{3+} > Ca^{2+} > Na^+ \text{ and } La^{3+} > Sr^{2+} > K^+ \quad (22)$$

In general, the most stable coordination bonds can cause smaller and highly charged rations to form more stable coordination compounds.

4.4 Electronegativity

When an electron pair attracts a central ion toward itself, a strong stability complex is formed, and this is due to electron donation from ligand → metal ion. This donation process is increasing the bond stability of metal complexes exerted the polarizing effect on certain metal ions. Li^+, Na^+, Mg^{2+}, Ca^{2+}, Al^{3+}, etc. are such type of metal cation which is not able to attract so strongly from a highly electronegative containing stable complexes, and these atoms are O, N, F, Au, Hg, Ag, Pd, Pt, and Pb. Such type of ligands that contains P, S, As, Br and I atom are formed stable complex because these accepts electron from M → π-bonding. Hg^{2+}, Pb^{2+}, Cd^{2+}, and Bi^{3+} metal ions are also electronegative ions which form insoluble salts of metal sulfide which are insoluble in aqueous medium.

4.5 Temperature and pressure

Volatile ligands may be lost at higher temperature. This is exemplified by the loss of water by hydrates and ammonia:

$$[Co(NH_3)_6]Cl_3 \ (\Delta 175 - 180°C) \rightarrow [Co(NH_3)_5Cl]Cl_2 + NH_3 \quad (23)$$

The transformation of certain coordination compounds from one to another is shown as follows:

$$AgHg[AgI_4] \text{ (red)}(45°C) \leftrightarrows Ag_2[HgI_4] \text{ (yellow)} \quad (24)$$

4.6 Ligand nature

A ligand is an ion or small molecule that binds to a metal atom (in chemistry) or to a biomolecule (in biochemistry) to form a complex, such as the iron-cyanide coordination complex Prussian blue or the iron-containing blood-protein hemoglobin. The ligands are arranged in spectrochemical series which are based on the order of their field strength. It is not possible to form the entire series by studying complexes with a single metal ion; the series has been developed by overlapping different sequences obtained from spectroscopic studies [19]. The order of common ligands according to their increasing ligand field strength is

$$\begin{aligned}&O_2^{2-} < I^- < Br^- < S^{2-} < SCN^- \text{ (S–bonded)} < Cl^- < N_3^- < F^- < NCO^- < OH^- < C_2O_4^{2} \\ &- < H_2O < NCS^- \text{ (N–bonded)} < CH_3CN < \text{gly (glycine)} < \text{py (pyridine)} \\ &< NH_3 < \text{en (ethylenediamine)} < \text{bipy (2, 2' – bipyridine)} \\ &< \text{phen (1, 10 – phenanthroline)} < NO_2^- < PPh_3 < CN^- < CO < CH_2\end{aligned} \tag{25}$$

The above spectrochemical series help us to for determination of strength of ligands. The left last ligand is as weaker ligand. These weaker ligand cannot forcible binding the 3d electron and resultant outer octahedral complexes formed. It is as- $Mn^{2+} < Ni^{2+} < Co^{2+} < Fe^{2+} < V^{2+} < Fe^{3+} < Cr^{3+} < V^{3+} < Co^{3+}$. For the given ligand, it is not possible to say about the exerted strong or weaker field on the central metal ion. The values of Δ are observed as:

1. Increasing the oxidation number the value of Δ increased.

2. Δ increases from top to bottom.

However, when we consider the metal ion, the following two useful trends are observed:

1. Δ increases with increasing oxidation number.

2. Δ increases down a group. For the determination of stability constant, the nature of the ligand plays an important role.

The following factors described the nature of ligands.

4.7 Size and charge

The size and charge are two factors that affect the production of metal complexes. The less charges and small sizes of ligands are more favorable for less stable bond formation with metal and ligand. But if this condition just opposite the product of metal and ligand will be a more stable compound. So, less nuclear charge and more size= less stable complex whereas if more nuclear charge and small in size= less stable complex. We take fluoride as an example because due to their smaller size than other halide and their highest electro negativity than the other halides formed more stable complexes. So, fluoride ion complexes are more stable than the other halides:

$$FeF_2^+ \ \log \beta = 6.0 \tag{26}$$

$$FeCl_2^+ \ \log \beta = 1.3 \tag{27}$$

As compared to S^{2-} ion, O_2^{2-} ions formed more stable complexes.

4.8 Basic character

It is suggested by Calvin and Wilson that the metal complexes will be more stable if the basic character or strength of ligands is higher. It means that the donating power of ligands to central metal ions is high [20].

It means that the donating power of ligands to central metal ions is high. In the case of complex formation of aliphatic diamines and aromatic diamines, the stable complex is formed by aliphatic diamines, while an unstable coordination complex is formed with aromatic diamines. So, from the above discussion, we find that the stability will be grater if the e-donation power is greater.

Thus it is clear that greater basic power of electron-donating species will form always a stable complex. NH_3, CN^-, and F^- behaved as ligands and formed stable complexes; on the other hand, these are more basic in nature.

4.9 Ligand concentration

We know that if the concentration of coordination group is higher, these coordination compounds will exist in the water as solution. It is noted that greater coordinating tendency show the water molecules than the coordinating group which is originally present. SCN^- (thiocynate) ions are present in higher concentration; with the Co^{2+} metal ion, it formed a blue-colored complex which is stable in state, but on dilution of water medium, a pink color is generated in place of blue, or blue color complex is destroyed by $[Co(H_2O)_6]^{2+}$, and now if we added further SCN^-, the pink color will not appear:

$$\underset{\text{Blue}}{\left[Co(SCN)_4\right]^2} - +H_2O \leftrightarrows \underset{\text{Pink}}{\left[Co(H_2O)_6\right]^{2+}} + 4SCN^- \tag{28}$$

Now it is clear that H_2O and SCN^- are in competition for the formation of Co(II) metal-containing complex compound. In the case of tetra-amine cupric sulfate metal complex, ammonia acts as a donor atom or ligand. If the concentration of NH_3 is lower in the reaction, copper hydroxide is formed but at higher concentration formed tetra-amine cupric sulfate as in the following reaction:

$$CuSO_4 + NH_4OH \rightarrow Cu(OH)_2 \text{ (Small quantity of ligand)} \tag{29}$$

$$CuSO_4 + NH_4OH \rightarrow Cu(OH)_2[Cu(NH_4)2SO_4.H_2O] \text{ (High concentration of ligand)} \tag{30}$$

4.10 Chelating effect

For a metal ion, chelating ligand is enhanced and affinity it and this is known as chelate effect and compared it with non-chelating and monodentate ligand or the multidentate ligand is acts as chelating agent. Ethylenediamine is a simple chelating agent (**Figure 1**).

Due to the bidentate nature of ethylenediamine, it forms two bonds with metal ion or central atom. Water forms a complex with Ni(II) metal ion, but due to its monodentate nature, it is not a chelating ligand (**Figures 2** and **3**).

The dentate cheater of ligand provides bonding strength to the metal ion or central atom, and as the number of dentate increased, the tightness also increased. This phenomenon is known as chelating effect, whereas the formation of metal complexes with these chelating ligands is called chelation:

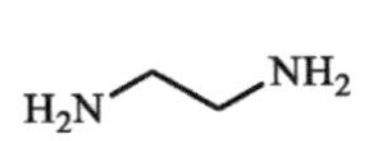

Figure 1.
Structure of ethylenediamine.

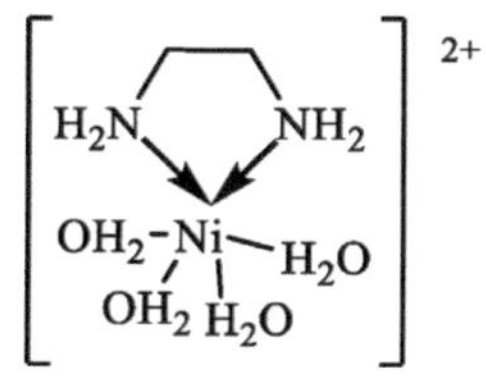

Figure 2.
Structure of chelating configuration of ethylenediamine ligand.

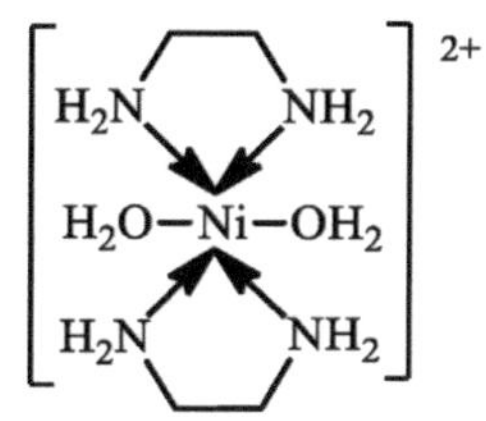

Figure 3.
Structure of chelate with three ethylenediamine ligands.

$$\text{Metal} + 2\ \text{Ligand} \leftrightarrow \text{MetalLigand}_2\ \text{K} = \frac{(ML2)}{[M][L]2} \tag{31}$$

$$\text{Metal} + \text{Ligand–Ligand} \leftrightarrow \text{MetalLigand} - \text{Ligand} \tag{32}$$

or 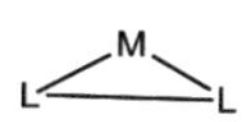 $$\text{K} = \frac{(ML-L)}{[M][L-L]} \tag{33}$$

Some factors are of much importance for chelation as follows.

4.11 Ring size

The sizes of the chelating ring are increased as well as the stability of metal complex decreased. According to Schwarzenbach, connecting bridges form the chelating rings. The elongated ring predominates when long bridges connect to the ligand to form a long ring. It is usually observed that an increased a chelate ring size leads to a decrease in complex stability.

He interpreted this statement. The entropy of complex will be change if the size of chelating ring is increased, i.e., second donor atom is allowed by the chelating ring. As the size of chelating ring increased, the stability should be increased with entropy effect. Four-membered ring compounds are unstable, whereas five-membered are more stable. So the chelating ring increased its size and the stability of the formed metal complexes.

4.12 Number of rings

The number of chelating rings also decides the stability of complexes. Non-chelating metal compounds are less stable than chelating compounds. These numbers increase the thermodynamic volume, and this is also known as an entropy term. In recent years ligands capable of occupying as many as six coordination positions on a single metal ion have been described. The studies on the formation constants of coordination compounds with these ligands have been reported.

The numbers of ligand or chelating agents are affecting the stability of metal complexes so as these numbers go up and down, the stability will also vary with it.

For the Ni(II) complexes with ethylenediamine as chelating agent, its log K_1 value is 7.9 and if chelating agents are trine and penten, then the log K_1 values are 7.9 and 19.3, respectively. If the metal ion change Zn is used in place of Ni (II), then the values of log K_1 for ethylenediamine, trine, and penten are 6.0, 12.1, and 16.2, respectively. The log βMY values of metal ions are given in **Table 1**.

$Ni(NH_3)_6^{2+}$ is an octahedral metal complex, and at 25 °C its log β_6 value is 8.3, but $Ni(ethylenediamine)_3^{2+}$ complex is also octahedral in geometry, with 18.4 as the value of log β_6. The calculated stability value of $Ni(ethylenediamine)_3^{2+}$ 10^{10} times is more stable because three rings are formed as chelating rings by ethylenediamine as compared to no such ring is formed. Ethylenediaminetetraacetate (EDTA) is a hexadentate ligand that usually formed stable metal complexes due to its chelating power.

4.13 Steric effect

A special effect in molecules is when the atoms occupy space. This is called steric effect. Energy is needed to bring these atoms closer to each other. These electrons run away from near atoms. There can be many ways of generating it. We know the repulsion between valence electrons as the steric effect which increases the energy of the current system [21]. Favorable or unfavorable any response is created.

For example, if the static effect is greater than that of a product in a metal complex formation process, then the static increase would favor this reaction. But if the case is opposite, the skepticism will be toward retardation.

This effect will mainly depend on the conformational states, and the minimum steric interaction theory can also be considered. The effect of secondary steric is seen on receptor binding produced by an alternative such as:

1. Reduced access to a critical group.

2. Stick barrier.

3. Electronic resonance substitution bond by repulsion.

4. Population of a conformer changes due to active shielding effect.

4.14 Macrocyclic effect

The macrocyclic effect is exactly like the image of the chelate effect. It means the principle of both is the same. But the macrocyclic effect suggests cyclic deformation of the ligand. Macrocyclic ligands are more tainted than chelating agents. Rather, their compounds are more stable due to their cyclically constrained constriction. It requires some entropy in the body to react with the metal ion. For example, for a

Metal ion	log βMY (25°C, I = 0.1 M)
Ca^{2+}	11.2
Cu^{2+}	19.8
Fe^{3+}	24.9

Table 1.
Metal ion vs. log βMY values.

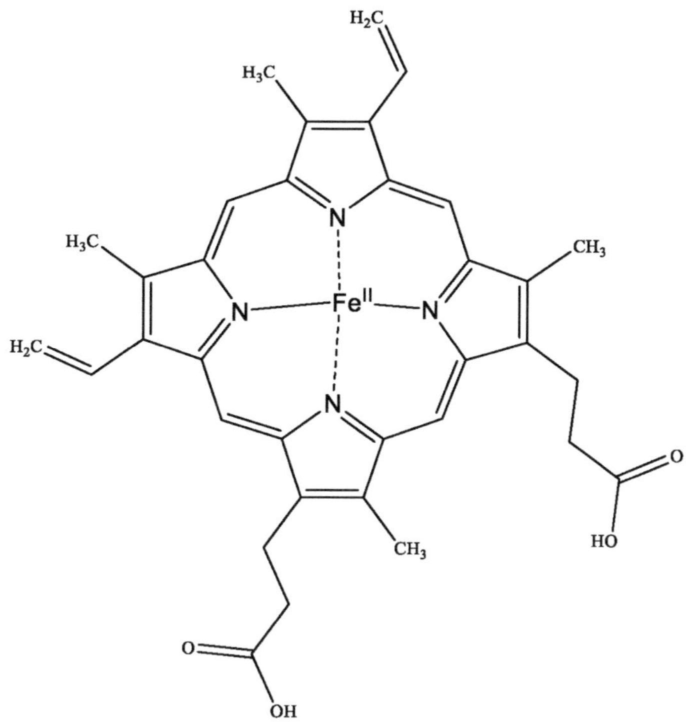

Figure 4.
Structure of hemoglobin is the biological complex compound which contains Fe(II) metal ion.

tetradentate cyclic ligand, we can use heme-B which forms a metal complex using Fe^{+2} ions in biological systems (**Figure 4**).

The n-dentate chelating agents play an important role for the formation of more stable metal complexes as compared to n-unidentate ligands. But the n-dentate macrocyclic ligand gives more stable environment in the metal complexes as compared to open-chain ligands. This change is very favorable for entropy (ΔS) and enthalpy (ΔH) change.

5. Determination of stability constants of complexes in solution

There are so many parameters to determination of formation constants or stability constant in solution for all types of chelating agents. These numerous parameters or techniques are refractive index, conductance, temperature, distribution coefficients, refractive index, nuclear magnetic resonance volume changes, and optical activity.

5.1 Methods based on study of heterogeneous equilibrium

5.1.1 Solubility methods

Solubility products are helpful and used for the insoluble salt that metal ions formed and complexes which are also formed by metal ions and are more soluble. The formation constant is observed in presence of donor atoms by measuring increased solubility.

5.1.2 Distribution method

To determine the solubility constant, it involves the distribution of the ligands or any complex species; metal ions are present in two immiscible solvents like water and carbon tetrachloride, benzene, etc.

5.1.3 Ion exchange method

In this method metal ions or ligands are present in solution and on exchanger. A solid polymers containing with positive and negative ions are ion exchange resins. These are insoluble in nature. This technique is helpful to determine the metal ions in resin phase, liquid phase, or even in radioactive metal. This method is also helpful to determine the polarizing effect of metal ions on the stability of ligands like Cu(II) and Zn(II) with amino acid complex formation.

5.1.4 Electrometric techniques

At the equilibrium free metal and ions are present in the solution, and using the different electrometric techniques as described determines its stability constant.

5.1.5 Potentiometric methods

This method is based upon the titration method or follows its principle. A stranded acid-base solution used as titrate and which is titrated, it may be strong base or strong acid follows as potentiometrically. The concentration of solution using 10^{3-} M does not decomposed during the reaction process, and this method is useful for protonated and nonprotonated ligands.

5.1.6 Polarographic method

This is the graphic method used to determine the stability constant in producing metal complex formation by plotting a polarograph between the absences of substances and the presence of substances. During the complex formation, the presence of metal ions produced a shift in the half-wave potential in the solution.

5.2 Other methods

5.2.1 Rate method

If a complex is relatively slow to form and also decomposes at measurable rate, it is possible, in favorable situations, to determine the equilibrium constant.

5.2.2 Freezing technique

This involves the study of the equilibrium constant of slow complex formation reactions. The use of tracer technique is extremely useful for determining the concentrations of dissociation products of the coordination compound.

5.2.3 Biological method

This method is based on the study of the effect of an equilibrium concentration of some ions on the function at a definite organ of a living organism. The equilibrium concentration of the ion studied may be determined by the action of this organ in systems with complex formation.

5.2.4 Spectrophotometric method

The solution of 25 ml is adopted by preparing at the 1.0×10^{-5} M ligand or 1.0×10^{-5} M concentration and 1.0×10^{-5} M for the metal ion: $\log K = \frac{[HL]}{[l]} + pH$

The solutions containing the metal ions were considered both at a pH sufficiently high to give almost complete complexation and at a pH value selected in order to obtain an equilibrium system of ligand and complexes.

In order to avoid modification of the spectral behavior of the ligand due to pH variations, it has been verified that the range of pH considered in all cases does not affect absorbance values. Use the collected pH values adopted for the determinations as well as selected wavelengths. The ionic strengths calculated from the composition of solutions allowed activity coefficient corrections. Absorbance values were determined at wavelengths in the range 430–700 nm, every 2 nm.

5.2.5 Bjerrum's method

For a successive metal complex formation, use this method. If ligand is protonate and the produced complex has maximum number of donate atoms of ligands, a selective light is absorbed by this complex, while for determination of stability constant, it is just known about the composition of formed species.

Bjerrum (1941) used the method stepwise addition of the ligands to coordination sphere for the formation of complex. So, complex metal–ligand$_{-n}$ forms as the following steps [22]. The equilibrium constants, K_1, K_2, K_3, ... K_n are called stepwise stability constants. The formation of the complex metal-ligand$_n$ may also be expressed by the following steps and equilibrium constants.

Where:

M = central metal cation

L = monodentate ligand

N = maximum coordination number for the metal ion M for the ligand

$$\text{Metal} + \text{Ligand} \leftrightarrows \text{Metal} - \text{Ligand}\ K_1 = \frac{(ML)}{[M][L]} \tag{34}$$

$$\text{Metal} - \text{Ligand} \leftrightarrows \text{Metal} - \text{Ligand}_2\ K_2 = \frac{(ML_2)}{[ML][L]} \tag{35}$$

$$\text{Metal} - \text{Ligand}_2 \leftrightarrows \text{Metal} - \text{Ligand}_3\ K_3 = \frac{(ML_3)}{[ML_2][L]} \tag{36}$$

$$\text{Thus Metal} - \text{Ligand}_{n-1} + \text{Ligand} \leftrightarrows \text{Metal} - \text{Ligand}_n\ K_n = \frac{(ML_n)}{[ML_{n-1}][L]} \tag{37}$$

5.2.6 Isotopic dilution method

If a complex ion is slow to reach equilibrium, it is often possible to apply the method of isotopic dilution to determine the equilibrium concentration of one or more of the species. Most often radioactive isotopes are used.

5.2.7 Conductance measurement method

This method was extensively used by Werner and others to study metal complexes. In the case of a series of complexes of Co(III) and Pt(IV), Werner assigned the correct formulae on the basis of their molar conductance values measured in freshly prepared dilute solutions. In some cases, the conductance of the solution increased with time due to a chemical change, e.g.,

$$[Co(NH_3)_4\,Br_2]^+ + 2H_2O\;[Co(NH_3)_4(H_2O)_2]^{3+} + 2Br^- \tag{38}$$

6. Conclusion

It is concluded that the information presented is very important to determine the stability constant of the ligand metal complexes. Some methods like spectrophotometric method, Bjerrum's method, distribution method, ion exchange method, electrometric techniques, and potentiometric method have a huge contribution in quantitative analysis by easily finding the stability constants of metal complexes in aqueous solutions.

Acknowledgements

All the authors thank the Library of University of Delhi for reference books, journals, etc. which helped us a lot in reviewing the chapter.

Author details

Jagvir Singh[1*], Abhay Nanda Srivastav[2], Netrapal Singh[3] and Anuradha Singh[4]

1 Department of Chemistry, ARSD College, University of Delhi, New Delhi, India

2 Department of Chemistry, Nitishwar Mahavidyalaya (BRABU), Muzaffarpur, Bihar, India

3 Department of Chemistry, Deen Dayal Upadhyaya University, Gorakhpur, UP, India

4 Department of Zoology, Raghuveer Singh Govt Degree College, Lalitpur, UP, India

*Address all correspondence to: singhjagvir0143@gmail.com

References

[1] Rossotti HS. Limitations of the ligand solubility method for studying complex formation. Journal of Inorganic and Nuclear Chemistry. Apr 1960;**13**(1-2): 18-21

[2] Jacqueline GG, Monica NL, Varinia LR, Juan ARV, Jose JN, Segoviano G. Spectrophotometric determination of the formation constants of Calcium(II) complexes with 1,2-ethylenediamine, 1,3-propanediamine and 1,4-butanediamine in acetonitrile. Journal of Green Energy & Environment (KeAi). 2017;**1**:51-57

[3] Syed AT, Feroza HW, Muhammad HSW, Saadia S, Allah NM, Allah BG. Spectrophotometric study of stability constants of cimetidine–Ni(II) complex at different temperatures. Arabian Journal of Chemistry. 2012;**2**: 309-314

[4] Thanavelan R, Ramalingam G, Manikandan G, Thanikachalam V. Stability constants of mixed ligand complexes of lead(II) with 1-(aminomethyl) cyclohexane acetic acid and α-amino acids. Journal of Saudi Chemical Society. 2014;**18**(3):227-233

[5] Abdelatty MR. Potentiometric studies on ternary complexes involving some divalent transition metal ions, gallic acid and biologically abundant aliphatic dicarboxylic acids in aqueous solutions, Beni-Suef University. Journal of Basic and Applied Sciences. 2015; **4**(2):174-182

[6] Katharina FP, Maria CB, Riccardo B, Bernard AG. Influence of pH on the speciation of copper(II) in reactions with the green tea polyphenols, epigallocatechin gallate and gallic acid. Journal of Inorganic Chemistry. 2012; **112**:10-16

[7] Vishenkova DA, Korotkova EI, Sokolova VA, Ratochvil BK. Electrochemical determination of some triphenylmethane dyes by means of voltammetry. Procedia Chemistry. 2015; **15**:109-114

[8] Lorenzo T, Zsolt B, Luca G, Attila F, Adrienn VMB. Thermodynamic stability, kinetic inertness and relaxometric properties of monoamide derivatives of lanthanide(III) DOTA complexes. Dalton Transactions. 2015;**44**:5467-5478

[9] Nagypal I. Chemistry of complex equilibria. Horwood. 1990;**85312**:143-145

[10] Dyrssen D, Ingri N, Sillen LG. Pit-mapping—A general approach to computer refinement of stability constants. Acta Chemica Scandinavica. 1961;**15**:694-696

[11] Ingri N, Sillen LG. High-speed computers as a supplement to graphical methods. Arkivor Kemi. 1964;**23**:97-121

[12] Sayce IG. Computer calculations of equilibrium constants of species present in mixtures of metal ions and complexing reagents. Talanta. 1968;**15**(12):1397-1421

[13] Sabatini A, Vacca A, Gans P. MINIQUAD—A general computer program for the computation of stability constants. Talanta. 1974;**21**(1):53-77

[14] Pearson RG. Chemical Hardness: Applications from Molecules to Solids. Manhattan, New York City: Springer-VCH; 2005. p. 210. ISBN: 978-3-527-60617-7

[15] Drago RS, Wong N, Bilgrien C, Vogel C. E and C parameters from Hammett substituent constants and use of E and C to understand cobalt-carbon bond energies. Inorganic Chemistry. 1987;**26**(1):9-14

[16] Vacca A, Nativi C, Cacciarini M, Pergoli R, Roelens S. A new tripodal

receptor for molecular recognition of monosaccharides. A paradigm for assessing glycoside binding affinities and selectivity by ^{1}H NMR spectroscopy. Journal of the American Chemical Society. 2004;**126**(50): 16456-16465

[17] Marcotte N, Taglietti A. Transition-metal-based chemo sensing ensembles: ATP sensing in physiological conditions. Supramolecular Chemistry. 2003;**15**(7): 617-717

[18] Boiocchi M, Bonizzoni M, Fabbrizzi L, Piovani G, Taglietti A. A di-metallic cage with a long ellipsoidal cavity for the fluorescent detection of dicarboxylate anions in water. Angewandte Chemie, International Edition. 2004;**43**(29):3847-3852

[19] Gampp M, Maeder M, Mayer CJ, Zuberbuhler AD. Calculation of equilibrium constants from multiwavelength spectroscopic data-I: Mathematical considerations. Talanta. 1985;**32**

[20] Frassineti C, Alderighi L, Gans P, Sabatini A, Vacca A, Ghelli S. Determination of protonation constants of some fluorinated polyamines by means of ^{13}C NMR data processed by the new computer program Hyp-NMR 2000. Protonation sequence in polyamines. Analytical and Bioanalytical Chemistry. 2003;**376**(7):1041-1052

[21] Jiaxin Z, Guoyu T, Peng S. Understanding thermodynamic and kinetic contributions in expanding the stability window of aqueous electrolytes. 2018;**4**(12):2872-2882

[22] Gans P, Sabatini A, Vacca A. Investigation of equilibria in solution. Determination of equilibrium constants with the HYPERQUAD suite of programs. Talanta. 1996;**43**(10): 1739-1753

Chapter 4

Chiral Mono- and α-Diimines and Their Pd(II) Complexes with Anticancer Activity

Guadalupe Hernández, Daniela Gutiérrez, Gloria E. Moreno, Oscar Portillo, René Gutiérrez and Eduardo Brambila

Abstract

The aim of this review is to provide mainly an outlook of the synthesis and characterization of chiral mono- and α-diimines ligands and their Pd(II) complexes carried out in our group in the last few years. Some other issues with simple chiral imines synthesized in our lab are also outlined. The report includes details about their versatile coordination patterns, biological activity in cancer cell lines, and engaging properties in different fields, such as materials science.

Keywords: chiral imines, Pd complexes, solvent-free reactions, anticancer activity

1. Introduction

The importance of Schiff bases resides in their structural variety as well as their ability to form a wide range of appealing structural arrangements depending of the constituents parent molecules with transition metals by acting as *N*-donor ligands, affording mono-, bi- and polynuclear complexes [1–3]. Accordingly, Schiff bases display a broad range of useful biological activities such as, *inter alia*, antibacterial, antifungal, antidiabetic, anti-inflammatory, and anticancer agents generating a huge interest in the medicine field [4–8]. The proper choice of the ligands in metal complex synthesis is essential for the activity that they could present since they determine some aspects like reactivity and lipophilicity.

We have focused our attention on the synthesis of chiral compounds since chirality is almost omnipresent in a broad range of organic molecules in the human body such as proteins, enzymes, amino acids, carbohydrates, and nucleosides. The body acts like a chiral selector metabolizing enantiomers by separate pathways and generating different pharmacological activities. For that reason, the current approach is to target specific molecules by designing more selective drugs, especially in chemotherapy where the distinction between cancerous and normal cells is essential for the success of the treatment and the reduction of the toxicity.

Likewise, the search of more eco-friendly procedures in the synthesis of organic molecules is one of the goals of our research group. Green Chemistry techniques like the use of microwave irradiation and solvent-free reactions display numerous advantages such as shorter reaction times, minimum waste, operational simplicity as well as reduction of thermal degradative byproducts along with cleaner work-up and generally higher yields [9, 10].

On the other hand, the discovery of anticancer activity of the cisplatin was a key event for the introduction of metal-based compounds to medicine, and the interest on these kind of compounds increased significantly in the last decades due to their ability to coordinate ligands in a three-dimensional configuration and bind to specific cell targets. Platinum-based drugs, particularly cisplatin, are widely use in the treatment of different types of cancer, but the toxicity and high resistance that they present limits their use. Therefore, the major challenge for chemists is the design of new drugs with less side effects. Efforts have been made to consider other metal-based complexes with cytotoxic properties, such as palladium complexes. They are known to show structural and thermodynamic analogy in regard to Pt(II) complexes, and display versatile coordination behavior and interesting properties. Palladium complexes of various donor-atom ligands have been found to possess engaging anti-tumor activity, as well as anti-inflammatory, anti-microbial, antiviral and antifungal properties [11, 12].

2. Chiral Pd(II) complexes

The incorporation of optically pure aromatic amines into α-dicarbonylic compounds bearing aromatic rings such as benzil in a 1/1 ratio generating enantiopure α-ketoimines was the first step for our investigations, considering that a flexible X=C—C=N (X = O, N) skeleton could lead to diverse coordination modes [13]. Then, the chiral mono-imine derived from (*S*)-(−)-1-phenylethylamine and benzyl under microwave radiation in solvent-free conditions led to the formation of the *N*-donor ligand (*S*)-(−)-(1-phenylethylimino)benzylphenylketone **1** which was allowed to react with K_2PdCl_4 giving two Pd complexes: a mono- **2** and a dinuclear Pd(II) **3** complexes (**Figure 1**).

On the other hand, *in vitro* assays are essential to determine the capacity of the compounds to modify basic cellular functions on different cancer cells. We have

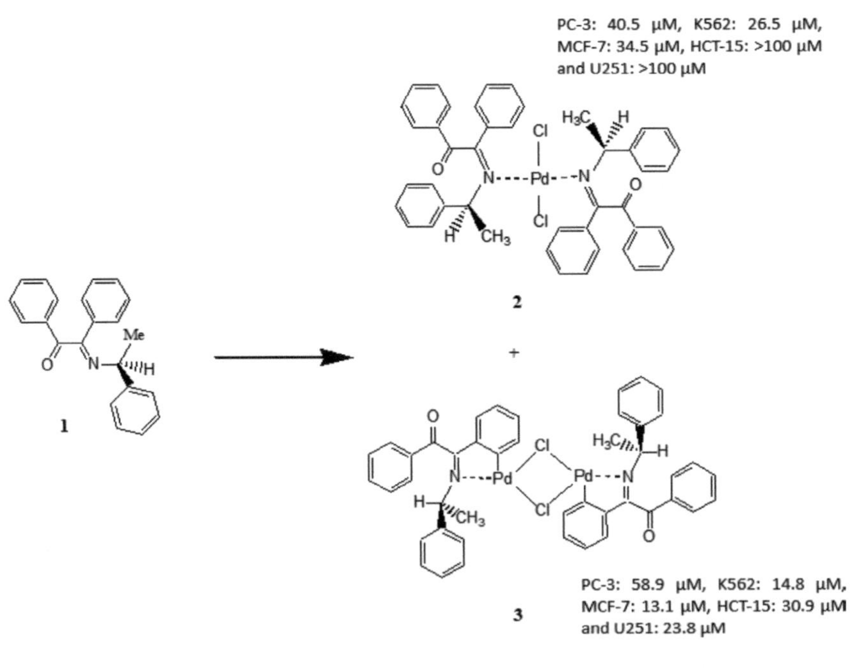

Figure 1.
Mono- and dinuclear chiral Pd(II) complexes **2** *and* **3** *with their respective* IC_{50} *values.*

employed sulforhodamine B staining to determine the cytotoxicity of our complexes, given the ease and high reproducibility of this method. By keeping constant the panel of human cancer cell lines (U251: central nervous system, PC-3: prostate cancer, K562: leukemia, HCT: colon cancer and MCF-7: breast cancer) we are able to compare the effects of the compounds in each cell and determine how the variations on the structure affect the activity. Those cell lines represent the most common types of cancer [14].

Complex **2** presented a common square planar geometry at the metal center with two Cl atoms *trans* and two ligands bonded through the N atoms N8 and N58 in a *trans* configuration while complex **3** is a dinuclear Pd(II) complex with one molecule in the asymmetric unit. For the binuclear complex, the coordination is carried through the N atoms (N8 and N58), like complex **2**, and the C atoms of the phenyl rings of the imino functions (C14 and C64). The low level of electronic delocalization in the ligand induced a high level of flexibility in the formation of complex **3**, producing a major distortion, due to the bite angles Cl-Pd-Cl and N-Pd-Cl.

The complexes **2** and **3** were tested by sulforhodamine B assays against U251, PC-3, K562, HCT and MCF-7 human cancer cell lines. Both compounds displayed cytotoxic activity, especially toward K562 (IC_{50}: 26.5 ± 0.4 and 14.8 ± 1.1 μM for complex **2** and **3**, respectively) and MCF-7 (IC_{50}: 34.5 ± 2.5and 13.1 ± 1.0 μM for complex **2** and **3**, respectively). In general, the binuclear complex was slightly better for all cell lines exhibiting lower IC_{50} values, while complex **2** surpassed the dose of 100 μM in U251 and HCT-15 cell lines [14].

Also, we have reported the synthesis of cyclopalladated compounds. Considering that our previous compounds displayed attractive properties, we decided to vary the substituents, replacing the aromatic rings in the α-dicarbonylic compounds by aliphatic substituents, such as two methyl groups and attaching also two chiral entities, i.e., to prepare α-diimines, as such kind of compounds have also a flexible N=C—C=N skeleton, displaying outstanding electron donor and acceptor properties and can potentially act in a variety of coordination modes. Then, the chiral diimines **4–5** were synthesized under solvent-free conditions starting from (*S*)-(−)-1-phenylethylamine and (*S*)-(−)-1-(4-methylphenyl) ethylamine with 2,3-butanedione, respectively. The reaction between Na_2PdCl_4 and each of the ligands **4–5** in a MeOH solution at ambient temperature led to the formation of the complexes **6** and 7 (**Figure 2**) [15]. In this case, the complex **6** is mononuclear

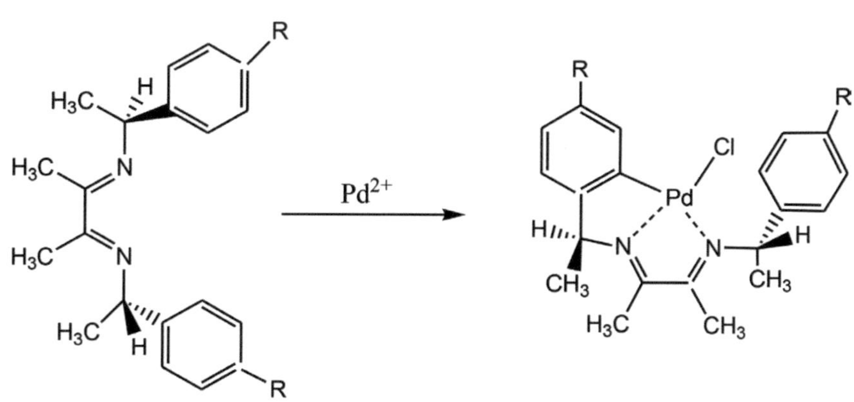

4 R= -H
5 R= -CH_3

6 R= -H
7 R= -CH_3

Figure 2.
Synthesis of chiral Pd complexes **6–7**.

with the Pd(II) center adopting a distorted square planar Pd[N_2CCl] coordination geometry where one benzene group is bonded to the metal center while the other is free of coordination. The steric hindrance produced by the benzene group and the formation of a Pd-C bound apparently blocked the dimerization of the complex. The solid obtained from complex **7** was not able to crystallize.

Both complexes **6** and **7** exhibited cytotoxic activity toward the panel of cultured cell lines previously mentioned, mainly against U251 and K562 cancer cells with IC_{50} values of 19.8 and 22.5 μM for complex **7**, and 23.6 and 25.44 μM for **6**, respectively. According to the data, **6** cannot be considered a good candidate as an anticancer agent since its IC_{50} values are too high for PC-3 and HCT-15, exceeding the dose of 100 μM. These compounds offer a better activity against U251 cell line compounds than the α-ketoimine complexes previously mentioned.

Thereafter, we carried out the synthesis of new unsymmetrical α-diimines by replacing one methyl group with a hydrogen atom and enlarging the number of chiral amines. A different method was used with the aim to improve the yields. Then, methylglyoxal and optically active aromatic and alicyclic primary amines were stirred in diethyl ether with Na_2SO_4 for 24 hours at room temperature leading to the formation of the ligands **8–11** (**Figure 3**). Solutions of the ligands **8–11** in benzene were treated this time with dichloro(1,5-cyclooctadiene) palladium (II) and stirred at room temperature under argon atmosphere to form complexes **12–15** (**Figure 4**). Worth-mentioning is that the coordination of the ligands took place in two different modes: chelating (σ, σ, N, N′) and monodentate (σ-N) [16].

Complexes **12** and **13** expose a *s-cis* chelate system and although they are chemically similar they crystallize in different way, in two distinct space groups. We believe that the crystal symmetry modification is a consequence of the crystallization rather than small conformational variations. The complex **15** displays two diimine ligands which are coordinated to the metal center in a *trans* square planar geometry, and the same behavior is observed in complex **14**. The importance of the *trans*-geometry around the Pd center has been attributed to the comparatively higher cytotoxicity values as those for *cis*-isomers.

It seems that the small substituents on the imine N atoms facilitates the orientation toward σ, σ, N, N′ coordination mode, stabilizing the complex through the chelate effect, while the monodentate (σ-N) coordination mode is favored by sterically hindered systems.

The results of the cytotoxic assay showed that Pd complexes with monodentate (σ-N) coordination mode (**14** and **15**) displayed IC_{50} values >100 μM; these complexes were dismissed for further assays because the doses required to inhibit cell

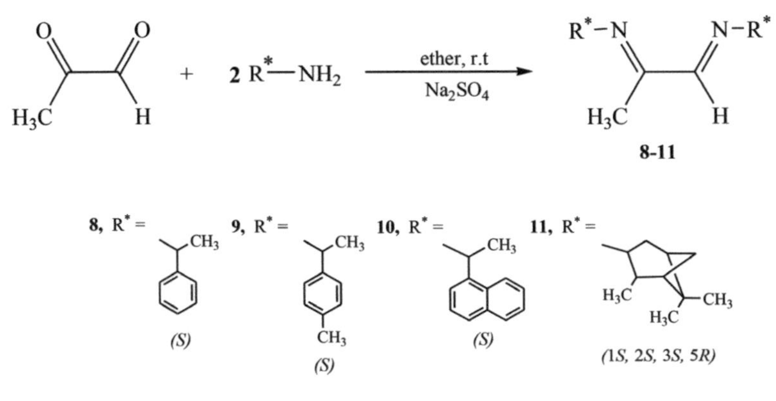

Figure 3.
Synthesis of chiral α-diimine ligands **8–11**.

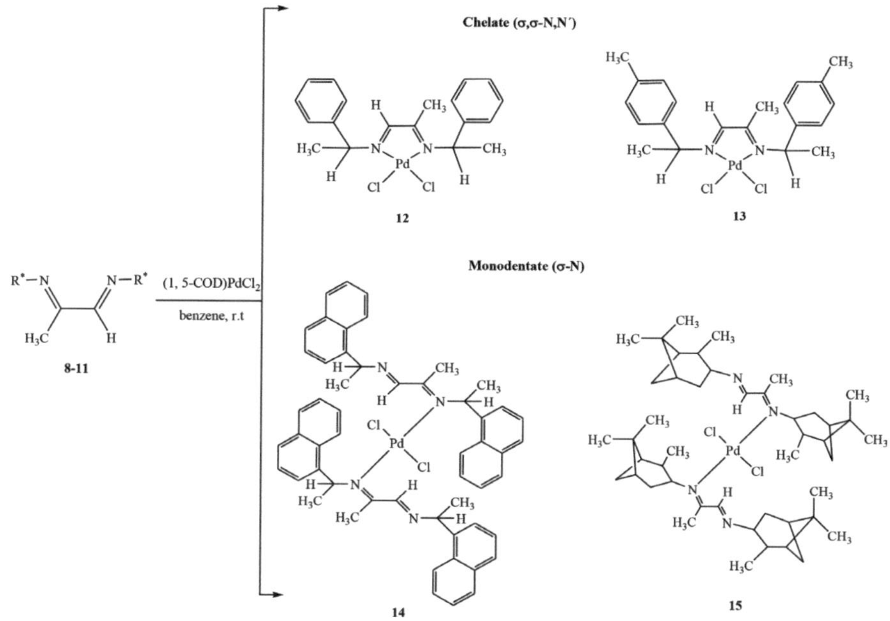

Figure 4.
Synthesis of chiral Pd(II) complexes **12–15**.

growth were too high. Complexes **12** and **13** also possessed cytotoxic activity against U251, PC-3, K562, HCT and MCF-7 cell lines, where IC_{50} ranged from 66 to 91 μM.

As such results were unpromising, we reconsidered the α-dicarbonylic compounds bearing aromatic rings, but this time with heterocyclic entities. By using the method previously used (microwave irradiation in solvent-free conditions), the chiral α-ketoimines **16–17** were synthesized from (S)-(−)-1-phenylethylamine and (S)-(−)-1-(4-methylphenyl) ethylamine with 2,2′-pyridil, respectively (**Figure 5**).

Complexes **18–19** (**Figure 6**) were synthesized by the reaction between $Pd(COD)Cl_2$ and each ligand **16-17** in a solution of benzene. It was not possible to obtain a monocrystal of complex **19**, however the crystal data of **18** showed that α-ketoimine **16** is a bidentate ligand and Pd(II) displayed a square-planar coordination geometry. In the case of **16**, the conjugation of imine and carbonyl double bonds with the aromatic systems and the substitution of vicinal C1 and C2 by pyridil rings implied that the ligand adopted a *gauche* conformation [17].

The data from the sulforhodamine B assay evidenced that none of the compounds possess cytotoxicity toward K562, however they are able to inhibit cell

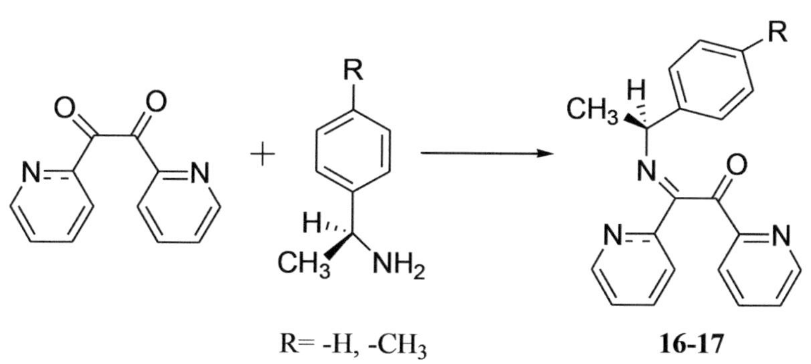

Figure 5.
Synthesis of chiral imines **16–17**.

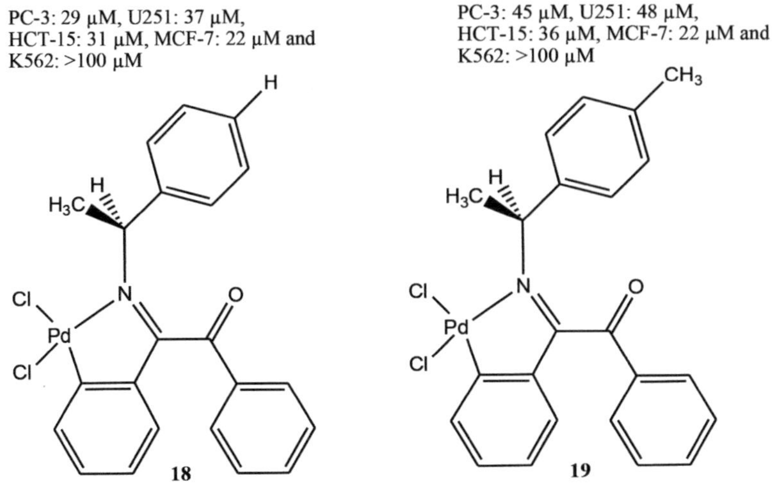

Figure 6.
Chiral Pd(II) complexes **18–19***.*

growth in U251, PC-3, HCT-15 and MCF-7, being **18** slightly better than **19** for all cell lines. The studies suggest that the nature of the aromatic rings have an impact in the cytotoxicity and the coordination mode.

Such results were not particularly impressive (at least a factor of 10 poorer than cisplatin), but they certainly do show variations in activity as well in the other cases.

It must be pointed out that even when the Pd-Schiff Base-complexes displayed cell growth inhibition against different classes of cancer, the IC_{50} that they have showed are not comparable with cisplatin. In general, Pd(II) complexes are kinetically less stable than those of Pt(II), by losing their structural integrity in biological fluids in a short period of time due to their rapid exchange. More specific studies *in vitro* and *in vivo* need to be done to determine their toxicity and to understand in a better way the mechanisms of action since it will aid the development of more efficient palladium-based drugs.

On the other hand, considering other alternatives to the flexible X=C—C=N (X = O, N) skeleton, for example as a heterodiene, we have also reported the microwave-assisted Diels-Alder [4+2] cycloaddition reaction of the optically pure α-ketoimines **20–21** and α-diimines **22–23**, with fullerene C_{60}. The chiral α-ketoimines **20–21** were readily synthesized in quantitative yield under solvent-free conditions starting from (*S*)-(−)-1-phenylethylamine and (*S*)-(−)-1-(4-methylphenyl) ethylamine with pyruvaldehyde, respectively, and upon reaction of C_{60} under focused-microwave irradiation in benzene, after 20 min the formation of the adducts **24–25** was observed (**Figure 7**) [18].

With the chiral α-diimines **22–23**, which were also readily prepared from (*S*)-(−)-1-phenylethylamine and (*S*)-(−)-1-(4-methylphenyl) ethylamine with pyruvaldehyde, respectively, the adducts **26–27** were obtained (**Figure 8**).

In addition, extending our studies to include some other transition metals, we have reported the preparation of chiral Hg(II) complexes with simpler chiral imines **28–30** as they present some relevant crystallographic features along with antimicrobial activity [19]. Thus, the solvent-free reaction of 2-pyridylcarboxaldehyde with optically active aromatic and alicyclic primary amines afforded the chiral imines **28–30** in almost quantitative yields (see **Figure 9**).

Solutions of the chiral imines **28–30** in methanol were treated with $HgCl_2$ with stirring at room temperature for 1 h, leading to the formation of complexes **31–33** (**Figure 10**).

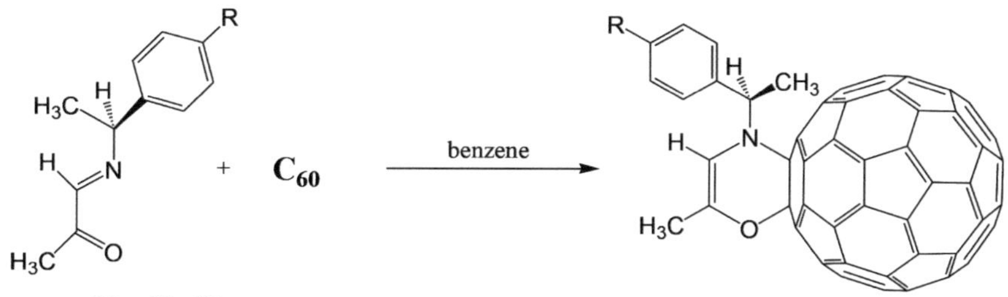

20 R=H
21 R=CH_3

24 R=H
25 R=CH_3

Figure 7.
Adducts **24–25** *synthesized.*

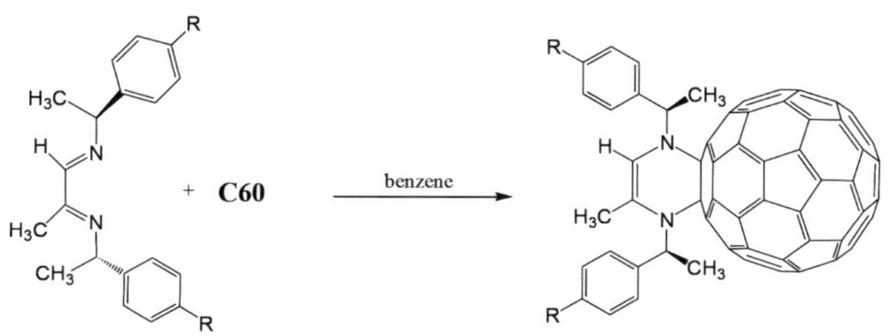

22 R=H
23 R=CH_3

26 R=H
27 R=CH_3

Figure 8.
Adducts **26–27** *synthesized.*

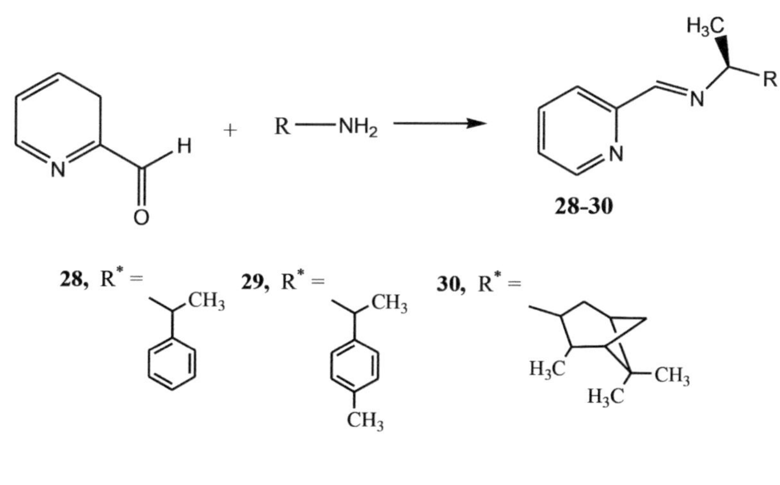

Figure 9.
Synthesis of chiral imines **28–30**.

Likewise, preliminary data have revealed that chiral imines **34–37** derived from 2-piridylcarboxaldehyde and the optically active aromatic amines (*S*)-(−)-1-(4-methylphenyl) ethylamine, (*S*)-(−)-1-(4-metoxyphenyl) ethylamine, (*S*)-(−)-1-(4-chlorophenyl) ethylamine and (*R*)-(+)-1-(4-fluorophenyl) ethylamine under solvent-free conditions (**Figure 11**) were allowed to react with $Zn(CLO)_4$ affording

31

32

33

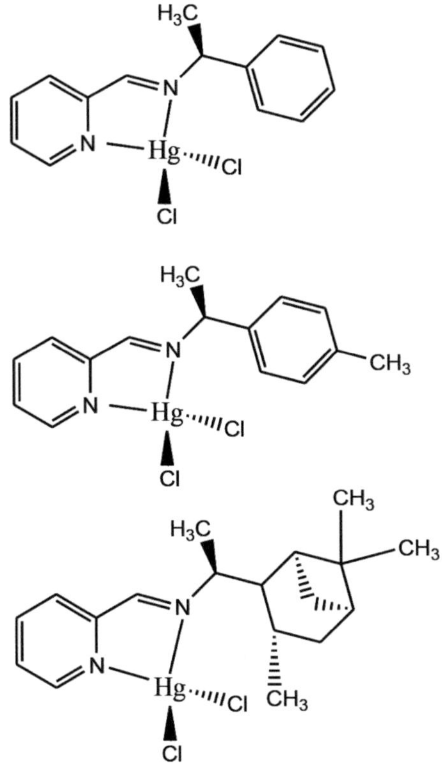

Figure 10.
Chiral Hg(II) complexes 31–33.

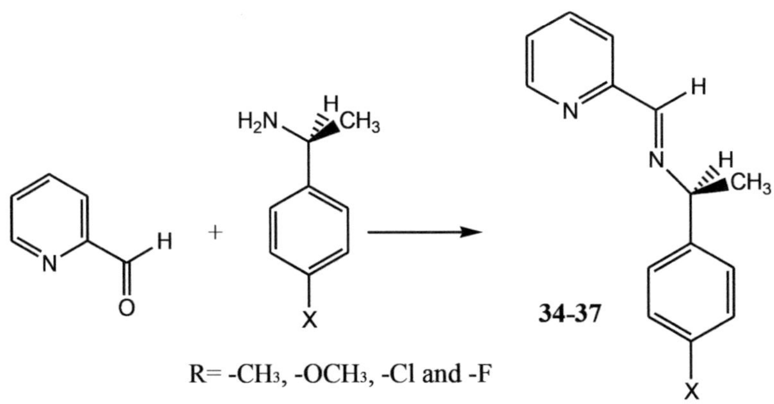

Figure 11.
Chiral imines 34–37.

Zn complexes **38–41** (**Figure 12**) with cytotoxic activity against the aforementioned human cancer cell lines as well as low toxicity in brine shrimps, along with antibacterial activity against *P. aeruginosa*, *E. coli* and *S. aureus*. Such results will be reported in due time.

On the other hand, simpler chiral imines have triggered interest in some other fields, especially in materials science; where by changing the substituents in the chiral moiety can afford morphological, optical and structural changes resulting in photoluminescent properties, which are extremely interesting since the viewpoint of physicists. In this context, we have recently reported a series of halogenated

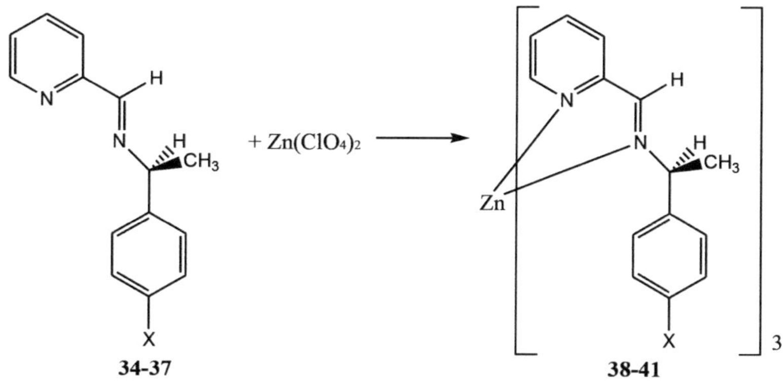

Figure 12.
Chiral Zn(II) complexes **38–41**.

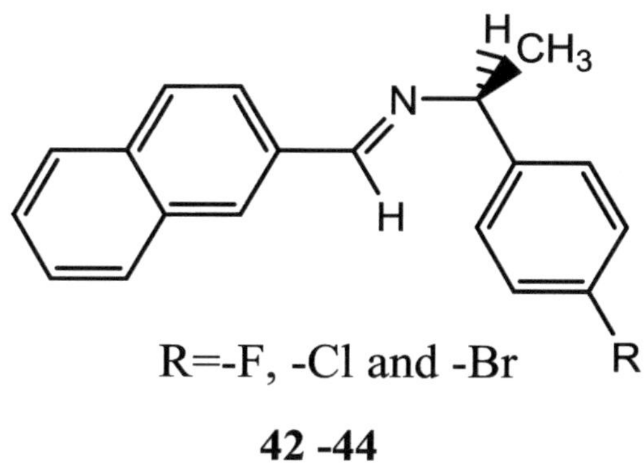

Figure 13.
Chiral halogenated imines **42–44**.

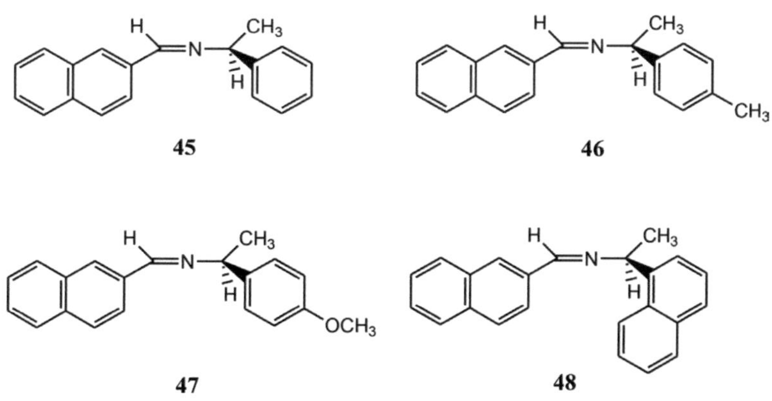

Figure 14.
Chiral imines **45–48**.

imines (**Figure 13**) derived from 2-naphtaldehyde and optically pure halogenated amines, under solvent-free conditions. As a result, imines **42–44** with a lamellar morphology exhibited photoluminescent properties. By changing the halogen atoms in the chiral moiety of the imines, the crystalline packing was modified. The bands observed in the visible region are caused by interstitial defects, vacancies, grain boundaries and stacking faults in the crystals The intensity of the bands increased in the following order: —F < —Cl < —Br, according to the increase in atomic radii. These features result attractive because of their possible applications in organic electroluminescent devices, organic light- emitting diodes, etc. [20, 21].

Likewise, in a series of chiral imines derived from 2-naphtaldehyde but with the halogen atoms in the *para*-position of the benzene ring of the amines replaced by other functional groups, such as $—CH_3$, $—OCH_3$ and naphthyl groups, imines **45–48** (**Figure 14**) exhibited green luminescence. As the previous results, the variations on the functional group as well as the molecular packing determined the morphological changes and consequently the luminescent properties of the imines [22].

3. Conclusion

The chemistry of Schiff bases and their transition metal complexes, especially Pd, is a field that is being noticed not only for their remarkable biological properties but also for their extensive applications in other fields. This area requires further studies to be carried out, and improvements in the permeability and transport are some of the factors to take into account in the design of these metal-based complexes.

Acknowledgements

Support from VIEP-UAP is acknowledged.

Conflict of interest

The authors declare no conflict of interest.

Author details

Guadalupe Hernández[1*], Daniela Gutiérrez[1], Gloria E. Moreno[1], Oscar Portillo[1], René Gutiérrez[1] and Eduardo Brambila[2]

1 Laboratorio de Síntesis de Complejos, Facultad de Ciencias Químicas, Universidad Autónoma de Puebla, Ciudad Universitaria, Puebla, Mexico

2 Laboratorio de Química Clínica, Facultad de Ciencias Químicas, Universidad Autónoma de Puebla, Ciudad Universitaria, Puebla, Mexico

*Address all correspondence to: guadalupe.hernandez@correo.buap.mx

References

[1] Qin W, Long S, Panunzio M, Biondi S. Schiff bases: A short survey on an evergreen chemistry tool. Molecules. 2013;**18**:12264-12289. DOI: 10.3390/molecules181012264

[2] Vigato PA, Tamburini S. The challenge of cyclic and acyclic Schiff bases and related derivatives. Coordination Chemistry Reviews. 2004;**248**:1717-2128. DOI: 10.1016/j.cct.2003.09.003

[3] Martin SF. Recent applications of imines as key intermediates in the synthesis of alkaloids and novel nitrogen heterocycles. Pure and Applied Chemistry. 2009;**81**:195-204. DOI: 10.1351/PAC-CON-08-07-03

[4] Kajal A, Bala S, Kamboj S, Sharma N, Saini V. Schiff bases: A versatile pharmacophore. Journal of Catalysis. 2013;**2013**:1-12. Art.ID 893512. DOI: 10.1155/2013/893512

[5] Zoubi WA, Al-Hamdani AAS, Kaseem M. Synthesis and antioxidant activities of Schiff bases and their complexes: A review. Applied Organometallic Chemistry. 2016;**30**:810-817. DOI: 10.1002/aoc.3506

[6] da Silva CM, da Silva DL, Modolo LV, Alves RB, de Resende MA, Martins CVB, et al. Schiff bases: A short review of their antimicrobial activities. Journal of Advanced Research. 2011;**2**:1-8. DOI: 10.1016/j.jare.2010.05.004

[7] Przybylski P, Huczynski A, Pyta K, Brzezinski B, Bartl F. Biological properties of Schiff bases and azo derivatives of phenols. Current Organic Chemistry. 2009;**13**:124-148. DOI: 10.2174/138527209787193774

[8] Abd El-Halim HF, Omar MM, Mohamed GG, El-Ela Sayed MA. Spectroscopic and biological activity studies on tridentate Schiff bases and their transition metal complexes. European Journal of Chemistry. 2011;**2**:178-188. DOI: 10.5155/eurjchem.2.2.178.188.240

[9] Andrade CKZ, Dar AR. Applying green processes and techniques to simplify reaction work-ups. Tetrahedron. 2016;**72**:7375-7391. DOI: 10.1016./j.tet.2016.09.055

[10] Sarkar A, Santra S, Kundu SK, Hajra A, Zyryanov GV, Chupakin ON, et al. A decade update on solvent and catalyst-free neat organic reactions: A step forward towards sustainability. Green Chemistry. 2016;**18**:4475-4525. DOI: 10.1039/c6gc01279e

[11] Kapdi AR, Fairlamb JS. Anti-cancer palladium complexes: A focus on PDX_2L_2 palladacycles and related complexes. Chemical Society Reviews. 2014;**43**:4751-4777. DOI: 10.1039/c4cs00063c

[12] Lazarevic T, Rilak A, Bugarcic ZD. Platinum, palladium, gold and ruthenium complexes as anticancer agents: Current clinical uses, cytotoxicity studies and future perspectives. European Journal of Medicinal Chemistry. 2017;**142**:8-31. DOI: 10.1016/ejmech.2017.04.007

[13] Peláez MA, Ramírez T, Martínez M, Sharma P, Álvarez C, Gutiérrez R. Synthesis, crystal structures and anticancer activity of the new chiral mono- and dinuclear palladium (II) complexes derived from (S)-(−)-(1-phenylethylimino)benzylphenylketone. Zeitschrift für Anorganische und Allgemeine Chemie. 2004;**630**:1489-1494. DOI: 10.1002/zaac.200400140

[14] Monks A, Scudiero D, Skehan P, Shoemaker R, Paul K, Vistica D, et al. Feasibility of a high-flux anticancer drug screen using a diverse panel of cultured human tumor cell lines.

Journal of the National Cancer Institute. 1991;**83**:757-766. DOI: 10.1093/jnci/83.11.757

[15] Cruz S, Bernès S, Sharma P, Vazquez R, Hernández T, Portillo R, et al. Synthesis, characterization and anticancer activity of new palladacycles derived from chiral α-diimines. Applied Organometallic Chemistry. 2010;**24**:8-11. DOI: 10.1002/aoc.1570

[16] Vázquez J, Bernès S, Sharma P, Pérez J, Hernández G, Tovar A, et al. Synthesis, characterization and anticancer activity of new chiral Pd (II)-complexes derived from unsymmetrical α-diimines ligands. Polyhedron. 2011;**30**:2514-2522. DOI: 10.1016/j.poly.2011.06.020

[17] Gutiérrez D, Bernès S, Hernández G, Portillo O, Moreno GE, Sharma M, et al. New chiral α-ketoimine-Pd(II) complexes and their anticancer activity. Journal of Coordination Chemistry. 2015;**68**:3805-3813. DOI: 10.1080/00958972.2015.1084618

[18] Tovar A, Peña U, Hernández G, Portillo R, Gutiérrez R. Microwave-assisted synthesis of new adducts from the Diels-Alder [4+2]-cycloaddition reaction of chiral a-oxo imines and a-diimines with fullerene C60. Synthesis. 2007:22-24. DOI: 10.1055/s-2006-958932

[19] Hernández G, Bernes S, Portillo O, Ruíz A, Moreno GE, Gutiérrez R. Crystal structures of three mercury(II) complexes [HgCL2L] where L is a bidentate chiral imine ligand. Acta Crystallographica Section E. 2015;**E71**:1462-1466. DOI: 10.1107/S2056989015020368

[20] Hernández-Téllez G, Moreno GE, Bernes S, Mendoza A, Portillo O, Sharma P, et al. Crystal structures of ten enantiopure Schiff bases bearing a naphthyl group. Acta Crystallographica Section E. 2016;**E72**:583-589. DOI: 10.1107/S2056989016004692

[21] Sosa A, Portillo O, Hernández G, Moreno GE, Juárez G, Chávez M, et al. Optical and morphological changes in the solid state of chiral imines bearing halogen substituents (F, Cl and Br). Material Letters. 2017;**189**:313-316. DOI: 10.1016/j.matlet.2016.11.053

[22] Portillo Moreno O, Meléndez Bustamante FJ, Chávez Portillo M, Moreno Morales GE, Hernández Téllez G, Sosa Sánchez A, et al. Photoluminescent green emission band induced by the systematic change of —CH3, —OCH3 and naphthyl groups in chiral imines. Crystal Growth & Design. 2018;**18**:660-668. DOI: 10.1021/acs.cgd.7b00851

Chapter 5

Stability of Vanadium Chalcone Complexes

Vikas Vilas Borge and Raju M. Patil

Abstract

The vanadyl(IV) complexes of substituted chalcones were prepared by refluxing vanadyl sulphate with different substituted chalcones in ethanoic medium. The chalcones were prepared with different aromatic aldehydes like benzaldehyde, hydroxy benzaldehyde, nitro benzaldehyde and chloro benzaldehyde. The synthesized Vanadium complexes were characterized by different spectral techniques. The IR spectral studies revealed that the chalcone derivatives are bidentate ligand. Magnetic studies, electron spin resonance and UV studies suggest that the complexes are in square pyramidal geometry. Conductance measurements suggest that all complexes are non-electrolyte in DMF. The thermal study explained the stability of complex and its decomposition. The synthesized ligand and metal complexes were screened for their antibacterial activity against *E. coli* and *Staphylococcus aureus* bacterial strains and for antifungal activity against *P. notatum*.

Keywords: vanadyl sulphate, chalcone ligands, complexes, spectral studies, TGA, magnetic study, antifungal and antibacterial study

1. Introduction

Chemically chalcones consist of open chain flavonoids in which the two aromatic ring joined by a three carbon. Chalcones are α, β-unsaturated ketones and contain a reactive keto-ethylenic group 2-hydroxychalcones and their heterocyclic analogues are also reported to form coordination complexes.

Chalcones are widely distributed in fruits, vegetables, tea, spices and also have applications in biological activities. These are abundant in edible plants and are considered to be the precursors of flavonoids and isoflavonoids [1]. Some chalcones has been employed for the gravimetric estimation of divalent ions of palladium, copper and nickel. Calcium determination by complexometric titration by using chalcone-metanil as indicator also has been reported. Chalcones possesses several biological activities with wide ranging. Chalcones use for the treatment of chronic diseases which involve inflammatory processes, for example, diabetes mellitus [2]. The study of chalcone derivatives has become of much interest in recent years on account of their antibacterial, antiviral activities [3]. Many chalcones based have been recognized in a number of screening assays for modulating important pathways or molecular targets in cancer [4, 5].

The coordination chemistry of oxovanadium(IV) (i.e. VO^{2+} or vanadyl ion) is more interesting and rather more important because of two main reasons. Firstly, the vanadyl complexes are finding more and more importance in biological systems. Secondly, the coordination number and geometry of this metal is highly ligand dependent. Moreover vanadyl ion is less toxic. In last year's research has been directed towards the synthesis of efficient bioactive compounds with low toxicity, in order to achieve this goal the type and the position of substituent into ligand were varied [6].

A wide variety of oxovanaduim(IV) complexes have been prepared and characterized by various physicochemical methods. Oxovanadium(IV) complexes have a square pyramidal geometry with apical oxygen and the vanadium atom lying above the plane defined by the donor atoms of the equatorial ligands. These square pyramidal complexes generally exhibit strong tendency to remain five coordinate [7, 8]. The coordination chemistry of vanadium(V) compounds is dominated by oxo complexes, containing the VO^{3+} or the VO^{2+} moiety [9]. The absorption band due to V=O stretching vibration of oxovanadium(IV) complexes is usually observed at a higher wavenumber compared to those of vanadate(V) complexes. However, the V=O stretching vibration is susceptible to a number of influences including electron donation from basal plane ligand atoms, solid-state effects, and coordination of additional molecules. Therefore, there has been considerable work done to assign the V=O stretching frequencies in oxovanadium(IV) compounds [10].

An increasing research interest of vanadium in coordination chemistry is not only due to its exhibiting a range of oxidation dates from +5 to −1 but also complexities exhibited by vanadium complexes and their industrial [11, 12], biological [13, 14] and medicinal [15, 16] applications. Vanadium complexes have been reported to have interesting antibacterial activities [17–20]. It has been reported that V(IV) complexes high effect on anticancer activity [21].

2. Preparation of substituted chalcone

The substituted chalcones prepared by stirring the equimolar concentration mixture of 2-hydroxy-4,5-dimethyl acetophenone (0.01 mol) and substituted aromatic benzaldehyde (0.01 mol) in 20 ml ethanol for 1 h in presence of 50% NaOH. The mixture stirred till completion of the reaction (progress of reaction checked by TLC). The crude mixture poured into ice water then acidified the product with 10% hydrochloric acid. The coloured compound formed was filtered, washed with water and dried. The compounds recrystallized from absolute ethanol (**Figure 1**).

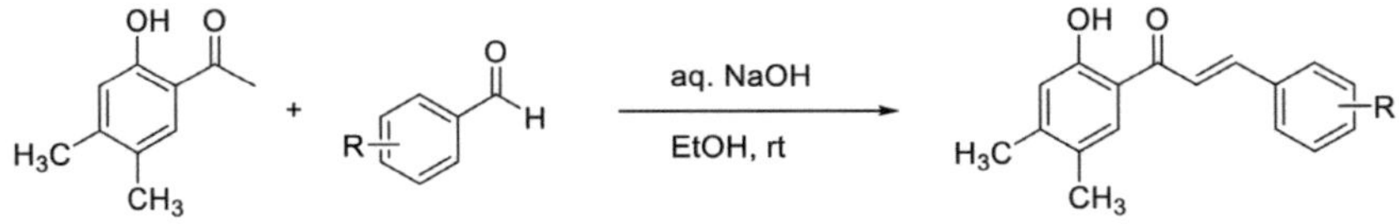

Figure 1.
Reaction scheme for chalcones (R = H, 4-OH, 4-NO$_2$ and 3-chloro).

3. Preparation of complexes

To a hot suspension of ligand (0.02 M) in ethanol, ethanolic solution (0.01 M) of the metal salt vanadyl sulphate ($VOSO_4$) was added drop wise with constant stirring and refluxed for 2–3 h. The resulting reaction mixture was cooled to room temperature and maintained up to pH 8.0 by adding ammonia then refluxed further for 30 min. The resultant product was filtered through Whatman filter paper no. 1 and repeatedly washed with ethanol until the washing were free from the excess of ligand. These complexes were finally dried under vacuum in desiccator over fused $CaCl_2$.

4. Instrumental methods

The IR spectra of complexes were recorded on a Perkin-Elmer instrument in KBr pallets in the range of 4000–400 cm^{-1}. TGA analysis of metal complexes were carried out in nitrogen atmosphere in the range 25–900°C on Rigaku Thermo Plus-8120 TG-DTA instrument with a heating rate 10°C min^{-1} using Alumina as a standard. UV-Visible spectra were recorded using DMF as solvent on Shimadzu UV-VIS spectrophotometer in the range 250–950 nm. Electron spin resonance spectra complexes were recorded on E-112 ESR Spectrometer as '*g*' marker ($g = 2.00277$) at room temperature. The conductance was measured in DMF solvent on Equiptronics Conductivity meter (EQ-664A).

5. Results and discussion

All the vanadium chalcone complexes are insoluble in water and most common organic solvents but sparingly soluble in DMF. The complexes are stable at room temperature. The elemental analysis shown in **Table 1** indicates that, all metal complexes have 1:2 (metal: ligand) stoichiometry for all the complexes (**Figure 2**).

$$VL_1{:}R_1{=}H \ \& \ R_2{=}H; VL_2{:}R_1{=}H \ \& \ R_2{=}OH$$

$$VL_3{:}R_1{=}H \ \& \ R_2{=}NO_2 \text{ and } VL_4{:}R_1{=}Cl \ \& \ R_2{=}H$$

All synthesized complexes are green in colour. All complexes are decomposed above 300°C. The molar conductance values obtained for these complexes at the concentration are in the range of 15–27 mhos.cm ^{2} mol^{-1} suggesting their non-electrolytic nature [22].

5.1 IR spectral studies

Some structurally significant IR bands for uncoordinated chalcones and complexes have been scanned in the 4000–400 cm^{-1} region and presented in **Table 2**. The complexes shows a band in the region 1162–1205 cm^{-1} which corresponds to

Complex	Empirical formula	Mol wt.	Yield %	% Elemental analysis found (calculated)						Molar conductance mhos.cm ^{2} mol^{-1}
				M	C	H	N	Cl	O	
V L_1	$C_{34}H_{30}O_5V$	569	62	9.10	71.52	5.15	—	—	14.13	18.21
				(8.96)	(71.70)	(5.27)			(14.06)	
V L_2	$C_{34}H_{30}O_7V$	601	59	8.55	67.94	4.91	—	—	18.58	23.46
				(8.48)	(67.88)	(4.99)			(18.63)	
VL_3	$C_{34}H_{28}N_2O_9V$	659	65	7.68	61.93	4.22	4.28	—	21.88	15.62
				(7.74)	(61.91)	(4.25)	(4.25)		(21.85)	
V L_4	$C_{34}H_{28}O_5Cl$ V	638	56	8.05	63.88	4.35	—	11.13	12.58	26.33
				(7.99)	(63.95)	(4.39)		(11.1)	(12.54)	

Table 1.
Physical and analytical data of metal complexes.

strong band in the region 1185–1221 in free ligand of chalcones due to υ (C—O). The shifts of this band were observed to move to lower wave numbers indicates bonding of phenolic oxygen atom to metal center via deprotonation. The band in the region 552–573 cm^{-1} can be assigned to the stretching modes of the metal to ligand bonds, ν (V—O) [23]. In addition the compound exhibit a strong band in the 952–983 cm^{-1} range due to the terminal V=O stretching and this value is close to the majority of oxovanadium(IV) complexes [24]. In the IR spectra of all the ligands an intense band appearing around 1640 cm^{-1} is attributed [25] to (C=O) group. In the complexes this band is observed in the 1618–1626 cm^{-1} range. The shifting of band to lower wave number indicates coordination through oxygen of (C=O) group. The band appears around 740 cm^{-1} in chloro complex indicating presence of C—Cl bond. The occurrence of band around 1500 cm^{-1} in VL_3 complex is due to ν (C—NO_2) mode.

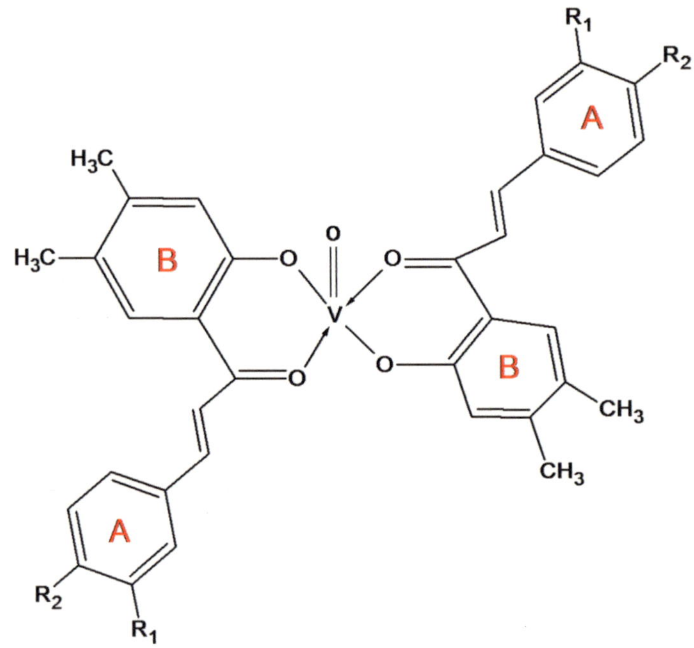

Figure 2.
Structure of vanadium complex.

Compound	OH group	C=O	C=C	C=O	C=CH(Aro)	V—O	—NO_2	C—Cl	V=O
L_1	2921	1640	1503	1221	2858	—			—
VL_1	3029	1626	—	1198	2964	552	—	—	963
L_2	3242	1639	1588	1221	2987	—			—
VL_2	3174	1621	1404	1205	3018	561	—	—	952
L_3	3390	1635	1513	1190	2921	—	1520		—
VL_3	3353	1623	1406	1168	3018	556	1490	—	983
L_4	3527	1635	1567	1185	2920	—		738	—
VL_4	3190	1618	1583	1162	3020	573	—	742	976

Table 2.
Important IR spectral bands (cm^{-1}) of the complexes.

5.2 Electronic spectra

Electronic absorption spectral study is an important tool used [26] to determine the geometry around a central metal atom. The electronic spectra (**Figure 3**) of the metal complexes in DMF (10^{-3} M) were recorded on Shimadzu UV-VIS spectrophotometer in the range 200–950 nm. The Vanadium chalcone complexes show absorption band in 380–770 nm region (**Table 5**). Most of the VO(II) complexes are green in colour. The electronic absorption spectra of VO(II) complexes showed three bands at 13,015–13,305, 17,455–17,632 and 25,755–25,930 cm^{-1}, with 17,455–17,632 cm^{-1} representing 10 Dq. These bands are assigned to the transitions $^2B_2 \rightarrow$ [2]$E(d_{xy} \rightarrow d_{xz},d_{yz})$,$^2B_2 \rightarrow {}^2B_1$ ($d_{xy} \rightarrow dx^2$-y [2]), and $^2B_2 \rightarrow$ [2]A_1 ($d_{xy} \rightarrow d_z{}^2$) respectively. From these electronic absorption bands and paramagnetic nature, square pyramidal geometry suggested for VO(II) chalcone complexes. As these oxovanadium(IV) belong to the 3d^1 system, there is no interelectronic repulsion in the metal ion so B_0 value of V^{4+} is not available. Therefore calculation of Racah parameter is not possible for these vanadium complexes (**Table 3**).

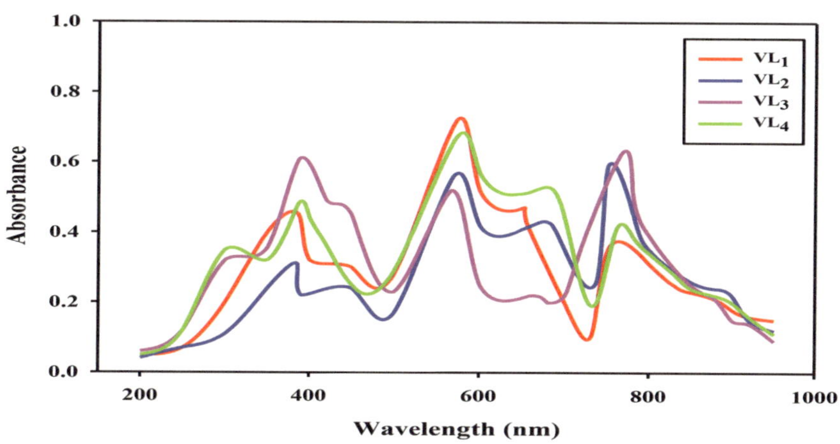

Figure 3.
Electronic spectra of V(IV) complexes.

Complexes	(υ1)	(υ2)	(υ3)	Dq (cm^{-1})	CFSE (KJ/mol)
VL_1	13,275	17,455	25,810	1745	−83.50
VL_2	13,305	17,585	25,930	1758	−84.12
VL_3	13,015	17,632	25,755	1763	−84.36
VL_4	13,135	17,490	25,890	1749	−83.69

Table 3.
Electronic spectral data for vanadium metal complexes.

Complex	Magnetic moment (B.M.)
VL_1	1.73
VL_2	1.71
VL_3	1.76
VL_4	1.72

Table 4.
Magnetic moment values of the vanadium (IV) chalcone complexes.

5.3 Magnetic susceptibility measurement

The magnetic susceptibility for all the Vanadium complexes at room temperature were recorded by the Gouy's method using $Hg[Co(SCN)_4]$ as a calibrant and reported in **Table 4**. The effective magnetic moments were calculated after applying diamagnetic corrections for the ligand components using Pascal's constants. The room temperature μ_{eff} values for the vanadium complexes were found in the range 1.71–1.76 B.M. The magnetic susceptibilities of the complexes are consistent with square pyramidal geometry around the central metal ion [27, 28]. The magnetic moments of the compounds investigated are in agreement with the findings of electronic absorption spectra.

5.4 Electron spin resonance study

ESR spectrum of VO(II) complexes was recorded at room temperature as polycrystalline sample, on *x*-band at 9.1 GHz, under the magnetic field strength 3000 G using TCNE as marker. The value of $g_{||}$ and $g_{\perp}$ is shown in **Table 5**. Paramagnetism of vanadyl ion arises from the single unpaired electron, its paramagnetism is due to spin angular momentum and $VO^{2+}$is most stable diatomic cation. In V(IV) complexes value of g is below value of free electron. The trend 2.0023 > $g_{\perp}$ > $g_{||}$, observed for the complexes. The EPR signal are split into eight lines suggested that V(IV) complexes are mononuclear [29]. The oxovanadium(IV) ion belong to the $3d^1$ system. Because of strong multiple covalent vanadium oxo-bond, a tetragonal compression occurs and therefore square pyramidal geometry proposed for VO(II) complex. The d orbital split into different level, dx^2-y^2 orbital is least stable as it is face to face with all four equatorial ligands. d_{xy} being most stabilized orbital therefore d^1 electron lies in d_{xy} orbital. As G values of present complexes are lower than four indicates, the ligands involve in the complex formation are the strong field ligands.

5.5 Thermal measurements

A plot of change in weight vs. temperature or time is known as Thermo gravimetric curve. It records the loss in weight as a function of temperature for transition that involves dehydration or decomposition. The TGA studies of the complexes was recorded in nitrogen atmosphere on Rigaku Thermo Plus-8120 instrument by increasing the temperature from room temperature upto 900°C at the heating rate of 10°C min^{-1}. The decomposition pattern obtained for VO(II) complexes of chalcone ligands follow two major stages (**Figure 4**). The thermal dehydration of this complex in the first stage was attributed to the loss of water leaving behind the nonhydrated complex takes place between 80 and 120°C. In the second stage the complex starts decomposing gradually which occurs in the temperature range corresponds to 220–310°C, this range correspond to the decomposition of part of chalcone ligand and the observed mass loss was recorded as 29.81% for VL_1, 32.41% for VL_2, 38.78%

Complex	$g_{\parallel}$	$g_{\perp}$	g_{av}	G axial symmetry parameter
VL_1	1.981	1.986	1.984	1.31
VL_2	1.958	1.983	1.975	2.29
VL_3	1.973	1.987	1.982	1.91
VL_4	1.949	1.987	1.974	3.48

Table 5.
ESR spectral data of vanadium(IV) metal complexes.

for VL_3 and 36.90% for VL_4 complex, which shows that at least unstable intermediate are also formed during this degradation stage. The third stage was decomposition of complete chalcone ligand in temperature range of 550–700°C. The overall total mass loss up to 700°C is around 88–90% which shows finally formation of VO. The residue left after 700°C is around 9–11% which resemble the theoretical values. Hence from thermal analysis it is clear that the complexes under study contains two water molecules for VL_1, VL_3 and VL_4 complex and three water molecules for VL_2 complex which are present in the ionic sphere of the complex (**Table 6**).

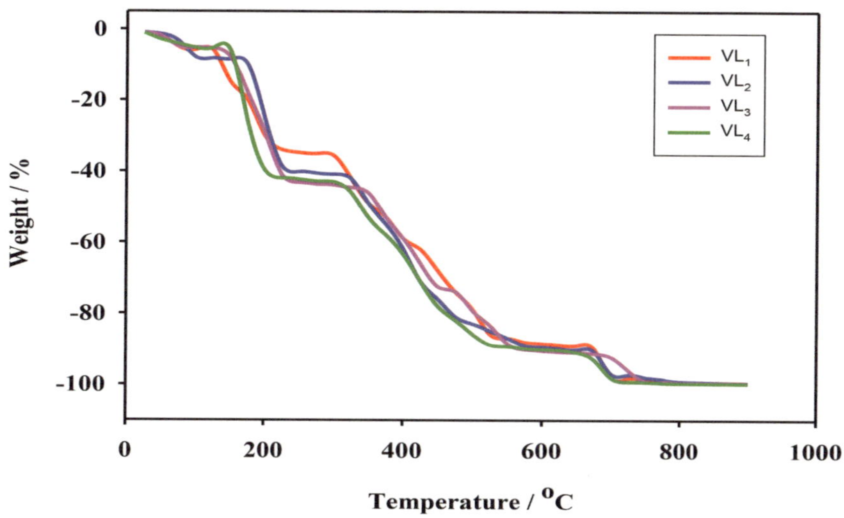

Figure 4.
Thermogravimetry spectra of V(IV) complexes.

Complex	Temperature range (°C)	% Weight loss		Decomposition product
		Found	Calculated	
VL_1	80–110	5.90	5.95	$2H_2O$
	235–290	29.81	29.75	$C_{14}H_{12}$
	550–680	53.26	53.20	$C_{20}H_{18}O_4$
	700–780	11.03	11.07	VO
VL_2	85–110	8.30	8.24	$3H_2O$
	230–300	32.41	32.36	$C_{14}H_{12}O_2$
	570–660	49.09	49.16	$C_{20}H_{18}O_4$
	700–750	10.16	10.22	VO
VL_3	80–115	5.12	5.17	$2H_2O$
	240–310	38.78	38.85	$C_{14}H_{10}O_4N_2$
	580–700	46.41	46.33	$C_{20}H_{18}O_4$
	720–780	9.70	9.64	VO
VL_4	90–120	5.27	5.34	$2H_2O$
	220–290	36.90	36.94	$C_{14}H_{10}Cl_2$
	550–660	47.85	47.77	$C_{20}H_{18}O_4$
	710–760	9.95	9.94	VO

Table 6.
Thermal data for vanadium metal complexes.

	Antibacterial activity		Antifungal activity
Compound	*E. coli*	*S. aureus*	*P. notatum*
	1.0 mg ml^{-1}	**1.0 mg ml^{-1}**	**250 μg disc^{-1}**
L_1	6	5	8
VL_1	9	8	9
L_2	8	9	11
VL_2	11	13	13
L_3	5	4	6
VL_3	7	6	7
L_4	7	8	10
VL_4	9	11	12
Neomycin	26	27	—

Table 7.
Antibacterial and antifungal activities (inhibition zone of bacterial growth, mm) of the ligand and metal complexes.

6. Antibacterial activity

It is known that killing of bacteria by chelated ligand is more powerful than non-chelated ligand, therefore chelated ligand acts as potent bactericidal agents. It was observed that due to delocalization of the π electrons over whole ligand and partial sharing of its positive charge with donor groups reduce the polarity of a metal ion in a complex. Thus, chelation increases the lipophilic character in complexes and results in an enhancement of activity [30, 31]. The antibacterial results, **Table 7**, showed that the complexes shows moderate activity against the bacteria *E. coli* and *S. aureus*.

7. Antifungal activity

Table 7 showed that metal complexes exhibited significant antifungal activity than chalcones at the same concentration against the fungi *P. notatum*. The order of inhibition with respect to metal complexes was $VL_2 > VL_4 > VL_1 > VL_3$.

It was concluded from above data that electron withdrawing substituents like nitro, cyno at para position in ring A decreases the antibacterial and antifungal activity. Electron donating group like hydroxyl, methoxy at para position increases the activity. Presence of halogen group at meta position in ring A shows good activity than the unsubstituted ring. The increased activity of the chelates can be explained based on the overtone concept and the Tweedy chelation theory [32]. According to the overtone concept of cell permeability, the lipid membrane that surrounds the cell favors the passage of only lipid-soluble material.

8. Conclusion

Based on the results the following conclusion maybe drawn. The higher decomposition temperature and electrical conductance studies show the presence of strong metal-ligand bonding and non-electrolytic nature of the complexes, respectively. Room temperature magnetic studies are indicative of paramagnetic nature and square pyramidal geometry of the VO(II) complexes which is supported by the electronic spectra. The IR spectra shows bonding of the metal through O-donor atoms of the ligands. In this study VO(II) complexes shows the moderate activity against *Escherichia coli*, *Staph aureus* bacteria and against *P. notatum* fungi.

Author details

Vikas Vilas Borge* and Raju M. Patil
Department of Chemistry, The Institute of Science, Mumbai, India

*Address all correspondence to: vikasborge@gmail.com

References

[1] Ahmad MR, Sastry VG, Bano N, Anwar S. Synthesis of novel chalcone derivatives by conventional and microwave irradiation methods and their pharmacological activities. Arabian Journal of Chemistry. 2016;**9**:931-935

[2] Najafian M, Ebrahim-Habibi A, Yaghmaei P, Parivar K, Larijani B. Core structure of flavonoids precursor as an antihyperglycemic and antihyperlipidemic agent: As in vivo study in rats. Acta biochimica Polonica. 2010;**57**:553

[3] Khan SA, Asiri AM. Arabian Journal of Chemistry. 2017;**10**:2890-2895

[4] Evangelista FCG, Bandeira MO, Silva GD, Silva MG, Andrade SN, Marques DR, et al. Anticancer activity of ruthenocenyl chalcones and their molecular docking studies. Medicinal Chemistry Research. 2017;**26**:27-43

[5] Khanapure S, Jagadale M, Bansode P, Prafulla C, Rashinkar G. Synthesis, spectroscopic and antimicrobial studies of oxovanadium (IV) complexes incorporating tridentate ONO donor hydrazones. Journal of Molecular Structure. 2018;**1173**:142-147

[6] Patil SK, Naik VM, Mallur NB. Comprehensive Coordination Chemistry. The synthesis, reactions, properties and applications of coordination compounds. Journal of Chemical and Pharmaceutical Research. 2012;**4**(4):2029-2036

[7] Wilkinson G, editor. Comprehensive Coordination Chemistry. Oxford: Pergamon Press; 1987

[8] Page EM, Wass SA. Vanadium 1995. Coordination Chemistry Reviews. 1997;**164**:203

[9] Nicholls D. In vitro study of the insulin-like action of vanadyl-pyrone and pyridinone complexes with a VO(O4) coordination mode. Coordination Chemistry Reviews. 1966;**1**:379-414

[10] Rangel M, Tamura A, Fukushima C, Sakurai H. Oxovanadium(V) Tetrathiacalix[4]arene complexes and their activity as oxidation catalysts. Journal of Biological Inorganic Chemistry. 2001;**6**:128

[11] Hoppe E, Limberg C. Vanadium catalysed asymmetric epoxidation of homoallylic alcohols. Chemistry: A European Journal. 2007;**13**:7006-7016

[12] Zhang W, Yamamoto H. Journal of the American Chemical Society. 2007;**129**:286-287

[13] Aureliano M, Gandara RMC. The role of vanadium in the chemical defense of the solitary tunicate, phallusia nigra. Journal of Inorganic Biochemistry. 2005;**99**:979-985

[14] Shobu O, Joseph P, Chem J. Insulin-Enhancing vanadium(III) complexes. Ecology. 2007;**33**:643-654

[15] Melchior M, Rettig SJ, Liboiron BD, Thompson KH, Yuen VG, McNeill JH, et al. Solid state and solution studies of vanadium(III)-L-cysteine compound and demonstration of its antimetastatic antioxidant and inhibition of neutral endopeptidase activities. Inorganic Chemistry. 2001;**40**:4686-4690

[16] Papaioannou A, Manos M, Karkabounas S, Liasko R, Evangelou AM, Correia I, et al. Synthesis and reactivity in inorganic, metal-Organic, and nano-metal chemistry microwave assisted synthesis and antimicrobial study of schiff base vanadium(IV) complexes of phenyl esters of amino acids. Journal of Inorganic Biochemistry. 2004;**98**:959-968

[17] Wazalwar SS, Bhave NS, Dikundwar AG, Ali P. Synthesis and Reactivity in Inorganic Metal-Organic and Nano-Metal Chemistry. 2011;**41**:459

[18] Liu J-L, Sun M-H, Ma J-J. Metal based biologically active compounds: Design, synthesis and antibacterial/antifungal/cytotoxic properties of triazole-derived schiffs bases and their oxovanadium(IV) complexes. Synthesis and Reactivity in Inorganic Metal-Organic and Nano-Metal Chemistry. 2015;**45**:117

[19] Chohan ZH, Sumrra SH, Youssoufi MH, Hadda TB. Synthesis, crystal structures and antibacterial studies of oxidovanadium(IV) complexes of salen-type schiff base ligands derived from meso-1,2-diphenyl-1,2-ethylenediamine. European Journal of Medicinal Chemistry. 2010;**45**:2739

[20] Taheri O, Behzad M, Ghaffari A, Kubicki M, Dutkiewicz G, Bezaatpour A, et al. New ni(II) complexes involving symmetrical bidentate N,O-donor Schiff base ligands: Synthesis at ambient temperature, crystal structures, electrochemical study, antioxidant and cytotoxic activities. Transition Metal Chemistry. 2014;**39**:253

[21] Abbasi Z, Salehi M, Kubicki M, Khaleghian A. Journal of Coordination Chemistry. 2017;**70**:2074

[22] Geasy MT. Spectroscopic studies of oxovanadium coordination compounds. Coordination Chemistry Reviews. 1971;7:81

[23] Baran EJ. Journal of Coordination Chemistry. 2001;**54**:215-238

[24] Syamal A, Kale KS. Inorganic Chemistry. 1979;**18**:992-995

[25] Dhar DN, Gupta VP. Indian Journal of Chemistry. 1971;**9**:818

[26] Cotton FA, Wilkinson G. Advanced Inorganic Chemistry. 5th ed. New York: John-Wiley & Sons; 1988

[27] Selwood PW. Magnetochemistry. 2nd ed. New York: Interscience; 1956

[28] Ballhausen CJ. Introduction to Ligand Field Theory. New York: McGraw-Hill; 1962

[29] Mustafi D, Makinen MW. Inorganic Chemistry. 2005;44:5580-5590

[30] Shrivastava RS. Metal-based antibacterial and antifungal amino acid derived Schiff bases: Their synthesis, characterization and in vitrobiological activity. Inorganica Chimica Acta. 1981;**56**:165

[31] Chohan ZH, Arif M, Sarfaz M. Synthesis, structural characterization and electrochemical and antibacterial studies of schiff base copper complexes. Applied Organometallic Chemistry. 2007;**21**:294

[32] Raman N, Kulandaisamy A, Tungaraja C, Manishankar P, Viswanthan S, Vedhi S. Transition Metal Chemistry. 2004;**29**:129

Chapter 6

Schiff Bases and Their Metallic Derivatives: Highly Versatile Molecules with Biological and Abiological Perspective

Muhammad I. Khan, Saima Gul and Murad Ali Khan

Abstract

1998 onwards, a span reporting thousands of research articles describes the ever-increasing applicability of Schiff bases and their metallic complexes; this chapter comprehensively examines the literature of the last 20 years. The structural diversity of these molecules made them available for a very wide range of biological and abiological applications. Schiff bases are excellent chelators and due to this unique property have found their place in qualitative and quantitative determination of metals in aqueous media. The structural diversity of metal chelates proved these to be outstanding catalysts and displayed interesting fluorescence effect. Finally, Schiff base moieties have found a unique position during the in vitro and in vivo experiments for drug development against a huge number of biological entities including bacteria, fungi, cancer cells, viruses, parasites, etc.

Keywords: synthesis, isomerization, electronic and vibrational spectra, catalyst, antitumor

1. Introduction

Even Hugo Schiff might not know while publishing his research findings as a scientific communication some 155 years ago that these organic compounds would become an independent area of immense interest not only in chemistry but also in other areas of scientific domains including physical, materials, biological, and engineering [1, 2]. Simply, Schiff bases can be defined as the products of chemical reaction of a primary amine with an aldehyde or a ketone under a given set of conditions. Structurally, oxygen of the carbonyl functionality (C=O) in an aldehyde and/or a ketone is replaced with nitrogen leading to the formation of imine or azomethine or anilino or azimethine (C=N) functionality with the liberation of a water molecule which is characteristic of Schiff bases [3]. There has been a substantial increase to study the coordination of the Schiff bases to the transition metals as well as main group metals and main group organometallics. The aforementioned metallic derivatives of Schiff bases have shown a rich diversity of structural motifs based upon coordinating ability of the ligand under investigation [4–7].

Schiff bases and their metallic derivatives have got outstanding applicability in biological systems acting as antibacterial, antifungal, antiviral, antitubercular, antitumor, insecticidal, bacteriostatic, in vitro cytotoxic, anti-inflammatory, analgesic, and antipyretic agents [8–15].

On the other hand, abiological applications of Schiff bases and their metallic derivatives are also very diverse, for example, in polymer chemistry, mechanochemical treatment often leads to the destruction of polymer under investigation, while very recently, the concept of mechanochemical polymerization has seen remarkable improvements which was initially reported with limited success [16]. Schiff base reaction is responsible for visual process in animals; the process starts with excitation of retinaldehyde leading to the formation of Schiff linkage with lysine and change in the membrane electrical potential with eventual transmission of signal to brain [17]. Schiff bases have been exploited very efficiently to monitor hazardous materials in the environment; for example, Cr^{3+} and organophosphates in the environmental samples were recognized, quantified, and removed with high accuracy and precision [18]. Electroanalytical techniques are very helpful for studying laboratory, clinical, and environmental samples as these are versatile and economical; Schiff bases have been known as ionophores and when fabricated with organic polymers can be transformed into membranes as ion-selective potentiometric sensors [19]. Schiff base ZnO complexes have been investigated as semiconductors by fabricating field-effect transistors to electronic performance [20]. Azo dye-based Schiff bases have been reported efficient chemosensors for detection and quantification of S^{2-} ions [21]. Photovoltaic characteristics of pyridine Schiff bases have been reported by illumination-dependent current–voltage measurements in solar cell applications [22]. Nickel Schiff base complexes have been studied for molecular docking experiments and their interaction with β-lactamase [23]. Schiff bases have successfully employed in vitro as well as in vivo for probing real-time sensing and analysis of Al^{3+} in a variety of diseases in human beings [24]. Polymeric Schiff bases strongly influence the electrochemical properties of the fuel cells and subsequent modification leading to an improved maximum power density in comparison to standard materials [25]. Metal–organic coordination polymers are widely used as conducting/semiconducting materials because of their accessible band gap 1–5 eV [26]. Schiff base palladium complexes immobilized on the mesoporous materials have been used as a heterogeneous catalyst for the Heck-Mizoroki coupling reaction and exhibited excellent catalytic activities for a wide range of alkenes [27, 28]. Schiff bases derived from cyclohexanediamine exhibited unusual structures, and these chiral molecules exhibited interesting photoluminescent properties [29].

2. Synthesis of Schiff bases and derivatives

The following routes are commonly adopted to prepare Schiff bases and their metallic derivatives:

- Reaction of aldehydes/ketones with primary amines under acidic/basic conditions [30]
- Aldol-like condensation of aldehyde [31]
- Oxidative synthesis of imines from alcohols and amines [32]
- Addition of organometallic reagents to cyanides [33]
- Reaction of phenols and phenol-ethers with nitriles [34]
- Reaction of metal amides [35]
- Synthesis of ketimines from ketals [36]

- Reaction of olefins and tertiary alcohols with hydrazoic acid [37]
- Conversion of α-amino acids into imines [38]
- Reduction of nitro compounds [39]

2.1 Isomerization of Schiff bases

Schiff bases tend to isomerize because of the imine group (C=N) to yield two stereoisomers *E* and *Z* (or *cis* and *trans*) isomers, and the formation of these stereoisomers can be controlled either kinetically or thermodynamically. Azobenzenes—common organic dyes—have been studied extensively owing to their potential applications in materials, medical, molecular switches, and other devices. Azobenzenes are photosensitive materials which undergo rapid reverse photoisomerization from the more stable *E* isomer to the less stable *Z* isomer. The exact mechanism of the *cis-trans* isomerization is not clear yet even after several decades of research. Reverse photoisomerization largely depends on the synthetic conditions like polarity of the medium; viscosity of the solvent used; and molecular substitutions on the azobenzene skeleton as these have very dramatic effects on the spectra of the molecules and the kinetics of isomerization [40].

In the literature, two possible mechanisms can be seen for the reversible photoisomerization of azobenzenes. First, the "inversion mechanism" (**Figure 1A**), proceeds via a linear transition state in which the N=N double bond remains undamaged, whereas the "rotation mechanism" (**Figure 1B**) occurs via a twisted transition state in which the N=N π-bond is broken. Extending the discussion, the trans form gets photoexcited, and an electron is excited from its ground state S0 orbital to its first or second singlet-excited state S1 or S2, retaining its spin under an n-π* or a π-π* excitation, respectively. Azo groups (N=N) photoisomerize via two distinct mechanisms: the π-π* transition with an out-of-plane rotation mechanism in which the N=N π bond is ruptured heterolytically or the n-π* electronic transition with an inversion of one sp^2 hybridized nitrogen atom through a *sp* hybridized linear transition state in which the double bond is retained. The rate of inversion isomerization is relatively

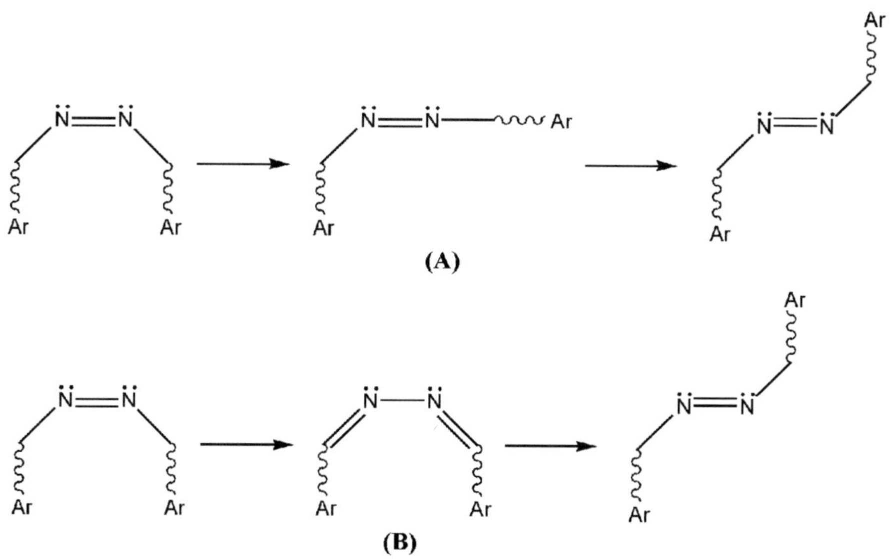

Figure 1.
Mechanisms of photoisomerization of azobenzenes.

rapid and mostly independent of the medium's polarity or the nature of substituents on the azobenzene, but the rate for the rotation mechanism increases rapidly with increasing solvent polarity. During the photochemical studies of the unique properties of azobenzenes, considering the polarity of solvent effect of six different solvents, namely, cyclohexane, toluene, benzene, tetrahydrofuran, acetone, and 3-pentanol, on the kinetics of the *cis-trans* isomerization of 4-anilino-4′-nitroazobenzene using a camera flash and a UV–Vis spectrophotometer. The data obtained revealed that the rate of *cis-trans* interconversion is solvent dependent, while solvent polarity has no effect on the rate of isomerization going through inversion mechanism. This strong relationship between rate of reactions and polarity of solvent pointed to an intermediate transition state that is considerably more polar than the *cis* conformation. The increase in polarity of solvent resulted in an obvious decrease in the activation energy, entropy, enthalpy, and Gibbs free energy of activation for the *cis-trans* isomerization process, while structural changes and nature of substituents also played an important role in the isomerization mechanisms of azobenzenes [41].

More recently, thermal *cis-trans* isomerization with detailed spectral and kinetic data of 4-aminoazobenzene has been examined in a range of solvents of with varying polarities. Interestingly, unlike azobenzene, the rate of thermal isomerization of 4-aminoazobenzene is highly dependent on solvent polarity with marked increase in rates in polarity of solvents when compared to nonpolar solvents. Moreover, inversion is the preferred pathway in *cis-trans* thermal isomerization in a nonpolar medium; but, under polar conditions, the isomerization adopted a rotational behavior. The same study concluded that kinetics and the mechanism of thermal isomerization is controlled by the polarity of the medium [42–44].

Recently, effect of fluorine atom as substituent at either side of the double bond has also been thoroughly examined in a thermodynamic and kinetic perspective of *cis-trans* isomerization reactions. The work also comprehensively explained *cis effect* produced by cis-oriented fluorine atoms on opposite ends of the double bond. This substitution resulted in the fluorine-*cis* (*Z*) isomer being lower in energy than the *trans* contrary to conventional wisdom, in which steric interaction between *cis* substituents lead to *trans* isomer lower in energy. The following is a summary of the results. For $CF_3CF{=}CHF$, the enthalpy of isomerization was measured from the equilibrium constant data as a function of temperature, and it was observed that the *Z* (*cis*) isomer was of lower energy than the *E* (*trans*) isomer, and "cis effect" of fluorine was very pronounced. In another case, *E* (*trans*) isomer was lower in energy than the *Z* (*cis*) because of the destabilizing steric interaction of the two relatively large groups. The measurement of the steric interaction between the CF_3 groups in the case of *Z*-$CF_3CH{=}CHCF_3$ allowed a better interpretation of the measured *E*(*trans*) to *Z*(*cis*) enthalpy of isomerization than that already reported for $CF_3CF{=}CFCF_3$. The absolute rate constant of 2-butene's *cis* to *trans* isomerization is also reported; the activation energies of the *E* isomers when compared with that of *E*-2-butene showed that the uncoupling energy of the π-bond decreased with fluorine substitution across the double bond [45].

3. Spectroscopic properties of Schiff bases

UV–Visible spectroscopy is a very useful analytical tool for studying the spectral properties of Schiff bases generally obtained as mixture of geometrical isomers, their sensitivity to the solvent properties, effect of substitution, pH, ambient temperature, etc. Absorbance spectra are generally recorded in the range of 300–450 nm; and the spectra are strongly solvent dependent and the corresponding *E* or *Z* isomer. With isatin Schiff bases under consideration, *Z*(*cis*) isomer appeared at higher wavelengths, that is, of lower energy when compared with *E*(*trans*) at lower wavelengths (higher energy); solvent-dependent electron density transfer was also observed, and the role of the position of substituent affected the appearance of frequency bands. Excited state was stabilized with a directional π-conjugative electron density shift in the molecule which might be a consequence of the long-range transmission of substituent effects, which supports the larger polarization of carbonyl group and thereby enhancing H-bonding capability of carbonyl oxygen and separation of electronic charges. Solvent affects are very complex phenomena, and the absorbance maxima altered inconsistently with respect to solvent and the position/type of the substituent(s) and geometrical features. In addition, other molecular properties like dipole moments, difference of electronegativity, partial charges, and chemical reactivity also significantly affected electronic spectra [46]. In the IR spectra, C=N is most commonly reported in the 1690–1640 cm^{-1} region as a strong and a sharp band at somewhat lower frequencies than the bands of C=O groups and close to C=C stretching frequencies. Angle strain, steric repulsion, and other complicated local factors and solution concentration and nature of solvent, in neutral solvents, the stretching frequency of C=N is found to be at 1670 cm^{-1}, and the corresponding force constant, 10.6 dynes cm^{-1}, is in the harmonic oscillator approximation. The frequency is usually lowered in the absence of one or more groups in conjugation with the C=N. Generally, there is no difference between IR and Raman frequencies and between the spectra of pure liquids and solids and their solutions in common organic solvents CCl_4 or other not very associative solvents. Generally, water of crystallization appeared at 3300 cm^{-1} plus range [46].

Multinuclear (^1H and ^{13}C) and multidimensional (HSQC, HMBC, COSY, and NOESY) NMR spectral analyses are helpful to establish absolute configuration of isatin Schiff bases. For example, the chemical shifts were assigned by a combined use of 1-D (^1H and ^{13}C proton-decoupled and 2-D NMR experiments COSY, NOESY, HSQC, and HMBC). 2-D NOESY experiments were used assess solution conformation of Schiff bases; *E* (*trans*) stereochemistry was assigned as major isomer in DMSO-d_6 solutions. Signals of some protons considerably shifted upfield by approximately 1 ppm relative to the same signals in the parent isatin; in contrast, chemical shifts of the protons present in *Z* (*cis*) isomer showed no difference to those found in parent isatin. Similarly, the same analogy was applied to understand to the electronic influences of substituents like —NO_2, and —OH and —OCH_3 group showed only the presence of *E* (*trans*) isomer. In contrast, electron-donating groups enhanced the stability of *Z* (*cis*) isomer; these factors affected the physico-chemical properties and lead to the differences in biological behaviors because of varying isomeric compositions [47].

Generally, in physical sciences and especially in chemistry, hydrogen bonding is one of the very important concepts for the deciding properties of the new materials. Intramolecular hydrogen bonds in ortho-hydroxyaryl Schiff bases, in ortho-hydroxyaryl ketones and amides, and in proton sponges and related compounds could be considered as classical objects for the investigation. The geometric and spectroscopic characteristics of the H-bonds that are typical for H-bonds are of medium strength, that is, it is shortened XY contact, elongated XH, directionality trend, change in frequency, and shielding and de-shielding. One of the most prominent aspects is the possibility of delocalization of proton leading to tautomers; and the existence of tautomer's equilibrium depends on room temperature, nature of solvent, and substitution. Proton delocalization is also observed due to the formation of stronger intramolecular H-bonds with larger proton potentials and due to thermally fluctuating media; either of these phenomena cause influence on the geometry of H-bond. Tautomeric equilibria changed with the change in the intrinsic geometry and spectroscopic properties of contributing tautomers, which further complicated the analysis of the experimental data. Computational work often provides the necessary insight via adiabatic and nonadiabatic PES calculations; and lastly, stabilization is also achieved due to delocalization of electrons in the chelate systems which are defined as resonance-assisted hydrogen bonds (RAHB) [48].

4. Schiff base metallic derivatives as catalysts

Schiff bases are easily supported on polymers and loaded with different metallic ions to check their catalytic action; and these polymer-supported catalysts are consistent in drastic reaction conditions like moisture and high temperature reactions. A careful survey of literature reveals that during the past two decades, several reports were seen on the synthesis of polymer-supported Schiff base complexes catalysts.

Transition metal Schiff base complexes are well-recognized homogeneous catalysts for various organic transformations with high homogeneity, good reproducibility, selectivity, and excellent catalytic activity to catalyze reactions under routine conditions. But there are certain drawbacks associated with these catalysts like corrosion, contamination of products, and separation of the catalysts. To address these challenges, heterogeneous catalysis is a good alternative approach; and there are two steps to achieve this task, that is, dispersing metallic ions on porous solid supports categorized as solid supported liquid phase catalyst (SLPC), and second step is the easy separation of the supported species from reaction mixture by filtration. Polystyrene is a well worked out cost-effective support, easily available, mechanically robust, chemically inert, and facile functionalization. When compared with the other supports like alumina or silica, the polymer-supported catalysts furnish the polymer chain flexibility for better microenvironment. Cobalt Schiff base complexes-functionalized polystyrene were effective and excellent catalysts in resolution of racemic mixtures. For the epoxidation of alkenes, early transition metal complexes proved excellent catalysts in combination H_2O_2; similarly, vanadium Schiff base complexes have been considered as versatile catalyst for oxidation of olefins, allylic alcohols, aromatic compounds, sulfides, and alcohols. There has always been a great need for high-quality pharmaceuticals, insecticides, and perfumes which significantly lead to develop enantioselective catalysts for epoxidation of olefins (**Figure 2**). More recently, chirally modified Li and Mg t-butyl peroxides have been used successfully in the epoxidation of electron deficient olefins. Henry's reaction also known as nitroaldol reaction is carbon–carbon

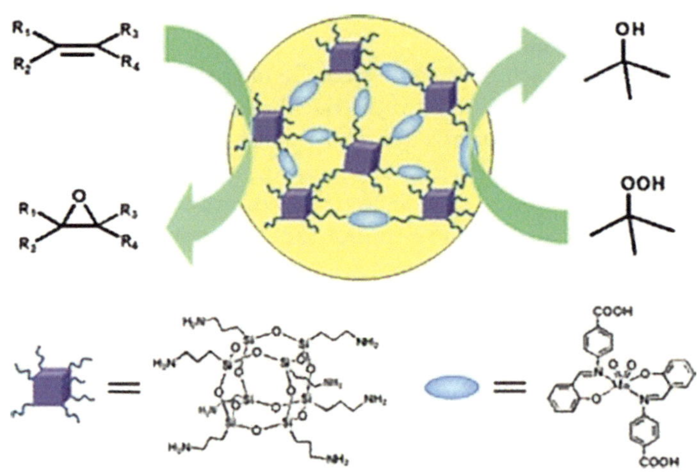

Figure 2.
Epoxidation of olefin using a polyhedral oligomeric silsesquioxane (POSS)-bridged oxo-molybdenum Schiff base complex [49–51].

bond coupling of carbonyl groups and nitro alkanes to generate organic intermediate natural products, drugs, and dyes. Schiff base Cu complex exhibited a very high enantioselectivity but limited to 2-nitrobenzaldehyde; also, Cu complexes of iminopyridines catalyzed reactions of nitromethane with aliphatic and aromatic aldehydes. Cu complexes of salen Schiff bases improved the chiral induction of Henry's reaction (yield: 95%); the m- and p-substituted benzaldehydes required lower temperatures and inert conditions [49–51].

5. Biological significance of Schiff bases and their metallic derivatives

The progress of biological inorganic chemistry has seen a marked interest in Schiff base complexes, as many of these complexes may be considered as models for biologically important species. Co, Ni, Fe, Zn, and UO_2 Schiff base complexes of 2-thiophene displayed good antibacterial activity against *E. coli*, *P. aeruginosa*, and *S. pyogenes*; these complexes also inhibited the growth of Gram-positive bacterial strains *S. pyogenes* and *P. aeruginosa*. This unique property could be applied safely in the treatment of infections caused by any of these strains [52].

Platinum complexes of salicylaldehyde and 2-furaldehyde Schiff base with o- and p-phenylenediamine were screened against *E. coli*, *B. subtilis*, *P. aeruginosa*, and *S. aureus*; the data proved the complexes were more potent than the parent Schiff bases. Complexes of Schiff bases derived from sulfametrole and varelaldehyde were screened against *E. coli* and *S. aureus*, and it was observed to have a significant effect on *E. coli*. Membrane of Gram-negative bacteria contains lipopolysaccharides; the reported Schiff bases and their metallic complexes could combine with these lipoic layers to enhance the membrane permeability of the Gram-negative bacteria. Lipophilicity is an important factor that controls the antibacterial activity as lipophilic cell membrane favored the passage of only lipid soluble materials; and this increase in lipophilic nature enhances the penetration of Schiff bases and their metallic complexes into the lipid membranes and seizing the growth of the organism. Schiff bases and their metallic derivatives are more toxic on *S. aureus* than on *E. coli*; it may be due to the SOH, OCH_3, and CH_3CH_2CH groups, interacting with the lipoic membrane (**Figure 3**) [53–55].

Cu, Ni, and Co complexes of 3-(2-hydroxy-3-ethoxybenzylideneamino)-5-methyl isoxazole and 3-(2-hydroxy-5-nitroben-zylidene amino)-5-methyl isoxazole Schiff bases were screened against *Aspergillus niger* and *Rhizoctonia solani*, and it was found that the activity increased upon coordination. The enhanced antifungal activity of the metal chelates can be explained based on chelation theory, which considers the overlapping of orbitals of each metallic ion with the ligand orbitals. Increased activity results in enhancement in the lipophilicities of the complexes due to delocalization of

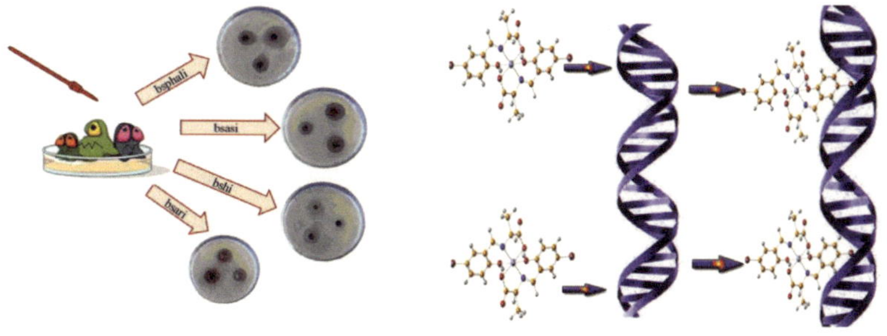

Figure 3.
Antibacterial activity and DNA interaction of Schiff bases [54, 55].

π-electrons in the chelate ring; it is also reported that increased lipophilicity leads to breakdown of the permeability barrier of the cell. Cu complexes proved to be more active, which may be attributed to higher stability constant than the other complexes [56–58].

Schiff base metal complexes conjugated with oligonucleotides were used to inhibit antitumor effect; zinc fingers are small protein motifs consisting of finger-like projections with a Zn^{2+} ion in each lump typically coordinated to two histidines and two cysteinic moieties. Binding with DNA takes place via classical (C_2H_2)-type zinc finger which regulates metathesis making these exciting targets for cancer therapy, especially breast, lung, liver, and brain cancers. The transition metal ions coordinated to Schiff bases have a very high affinity toward amino acid residues such as histidine and cysteine; their interaction with Zn fingers can, therefore, result in displacing Zn^{2+} from the Zn-finger and prohibiting the DNA binding with transcription factor. Co^{3+} Schiff base complexes were also attempted and showed greater specificity toward transcription factor inhibition than either of the precursors, that is, oligonucleotide and/or Co^{3+} complex, and Co^{3+} complex was required in >100-fold concentration for transcription factor inhibition, and the interaction with the protein was also non-specific [59–63].

Vaccination is a preemptive strategy to counter viral attacks like polio, small pox, chicken pox, etc. Viral attacks are a life-threatening situation for patients with compromised immunity, and a prompt treatment is required. To counter this, therapeutic options for viral infections are limited as antiviral agents in clinical practice are not yet completely effective, due to viral mutation against the drugs, and in addition, they also present several side effects. 1-amino-3-hydroxyguanidine salicylaldehyde Schiff bases are being considered as a good platform for the design of new and effective antiviral agents against mouse hepatitis with 50% growth inhibition at very low concentrations. Recently, abacavir-derived series of new Schiff bases prodrugs have been reported to use to treat human immunodeficiency virus (HIV) and acquired immunodeficiency syndrome (AIDS) under the trade name Ziagen (GlaxoSmithKline) [64].

Recently, researchers have focused work in search of metal-based antioxidants to identify the compounds having the potential in scavenging free radicals relevant to diseases associated with oxidative damage caused by reactive oxygen species. Synthetic antioxidants are widely used as effective and low-cost than natural antioxidants. Schiff base metallic derivatives have been reported as effective scavengers of reactive oxygen species acting as antioxidants [65].

6. Schiff base sensors

Sensor may be defined as system capable of detecting changes in its surroundings and transferring the information obtained to computer processor. During the past two decades, researchers have shown an increasing interest to develop new molecules/materials for the sensing of chemical species present in the environment. For example, inorganic anions and cations have a very huge impact on biological systems and the environment; these ions affect a wide range of chemical and biological processes in terms of toxicity to environment as well as human health. All the discussed points account for the aforesaid growth in the demand for ion-selective sensors. Anion sensors development is a very challenging task in comparison to cations as anions are larger in size, geometrically very diverse, and more importantly pH dependent. Chemosensors are abiotic molecules which have the selective and reversible binding ability with the analyte; azo Schiff bases have been reported as fluoride chemosensors. On the hand, Schiff base fluorescent sensors are becoming

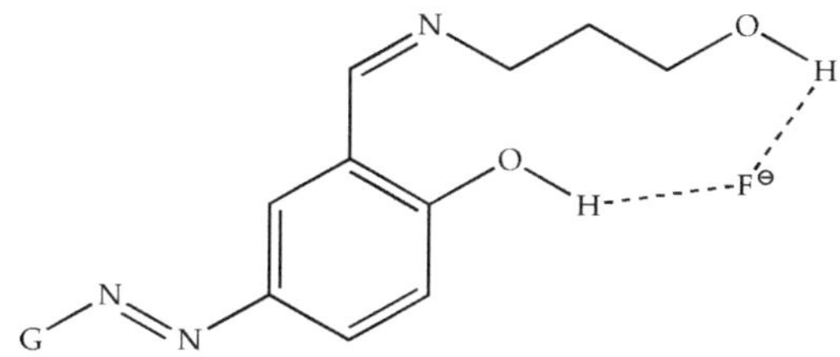

Figure 4.
Fluoride ion sensing by Schiff bases [66].

first choice as they exhibit high sensitivity even at very low concentration of analyte under consideration. In analytical chemistry, electrochemical methods are based on two fundamental parts, namely, *receptor*, and second is *transducer*, either performing transformation of the chemical information. The first does it into a form of energy, and second transforms the chemical information into a useful signal, respectively. Recently, azo Schiff bases chemosensors have been studied to show high selectivity and sensitivity for the detection of Hg^{2+}, Cd^{2+}, and fluoride ions in the presence of competing cations/anions; these chemosensors can be utilized in DMSO by both selective coloration and change in the absorption spectra with a more prominent color change that is observable to the naked eye (**Figure 4**) [66, 67].

7. Schiff base energy applications

Energy demand is a daunting challenge confronting the world as the quality of life depends largely on the availability of energy as we have burnt major part of fossil fuels. Renewable energy sources like solar light and wind generate larger amounts of energy and could potentially solve our immediate energy demands. Transformation of light into electric power in photovoltaic cells and subsequently light generation are two important interconnected phenomena which focus considerable interest to the researchers. Schiff base Zn^{2+} and Pt^{2+} complexes and organometallic compounds can produce, transform, and utilize energy due to their unique electronic setup which can be easily modified varying their chemical structure to develop a suitable material to fit an energy application. Remarkable work has been reported on salicylaldimine Schiff bases complexes with divalent Cu, Ni, Co, Zn, Mn, Fe, and Pt capable of being tailored into materials of applications in photophysics and optoelectronics [68].

Conflict of interest

The corresponding author on behalf of all the co-authors exclusively declares that there is/are no conflict(s) of interest(s) among authors to be declared.

Author details

Muhammad I. Khan*, Saima Gul and Murad Ali Khan
Department of Chemistry, Kohat University of Science and Technology,
Kohat, Khyber Pakhtunkhwa, Pakistan

*Address all correspondence to: gorikhan@kust.edu.pk; gorikhan@hotmail.com

References

[1] Schiff H. Disclosure from the University Laboratory Pisa: Examine the quinolone. Justus Liebigs Annalen der Chemie. 1864;**131**:112-117

[2] Schiff H. A new set of organic bases. Justus Liebigs Annalen der Chemie. 1864;**131**:118-119

[3] Furniss BA, Hannaford AJ, Smith PWG, Tatchell. VOGEL's Textbook of practical organic chemistry. 5th ed. Essex, UK: Longman Scientific & Technical; 1989

[4] Escuer A, Mayans J, Di Bari L, Font-Bardia M, Arrico L, Zinna F, et al. From Mesocates to Helicates: Nickel(II) Supramolecular Assemblies Derived from Tetradentate Schiff Bases. Structural, Magnetic and Chiro-Optical Studies. Chemistry—A European Journal. 2018 Mar 8. DOI: 10.1002/chem.201800323

[5] Lingjie H, Jinyin G, Xiangyu K, Chuan D, Shaomin S, Yu W. A novel highly sensitive fluorescent probe based on Schiff base for Al $^{3+}$. Scientia Sinica Chimica. 2017 Jul 18;**47**(8):1000-1006

[6] Silva CD, Silva DL, Modolo LV, Alves RB, Resende MD, Martins CV, et al. Schiff bases: A short review of their antimicrobial activities. Journal of Advanced Research. 2011;**2**(1):1-8

[7] Yao Y, Yang M, Zheng X, Tian L. Synthesis, characterization, and cytotoxic activity of triphenyltin complexes of N-(5-bromosalicylidene)-α-amino acids. Main Group Metal Chemistry. 2017 Aug 28;**40**(3-4):93-99

[8] Gopi C, Sastry VG, Dhanaraju MD. Microwave-assisted synthesis, structural activity relationship and biological activity of some new quinoxaline Schiff Base derivatives as highly potent spirochete bactericidal agents. Beni-Suef University Journal of Basic and Applied Sciences. 2017;**6**:39-47

[9] Shen X-H, Zhang Z-W, Shao L-J, Liu C. Synthesis, Crystal Structure, and Catalytic Property of a Copper(II) Complex Derived From 2-Hydroxy-5-Methoxybenzaldehyde Oxime. Synthesis and Reactivity in Inorganic, Metal-Organic, and Nano-Metal Chemistry. 2016;**46**:343-346

[10] Jarrahpour A, Khalili D, De Clercq E, Salmi C, Brunel JM. Synthesis, antibacterial, antifungal and antiviral activity evaluation of some new bis-Schiff bases of isatin and their derivatives. Molecules. 2007;**12**(8):1720-1730

[11] Qin W, Long S, Panunzio M, Biondi S. Schiff bases: A short survey on an evergreen chemistry tool. Molecules. 2013 Oct 8;**18**(10):12264-12289

[12] Zhu X, Wang C, Dang Y, Zhou H, Wu Z, Liu Z, et al. The Schiff base N-salicylidene-O, S-dimethylthiophosphorylimine and its metal complexes: synthesis, characterization and insecticidal activity studies. Synthesis and Reactivity in Inorganic and Metal-Organic Chemistry. 2000;**30**(4):625-636

[13] Bagihalli GB, Avaji PG, Patil SA, Badami PS. Synthesis, spectral characterization, in vitro antibacterial, antifungal and cytotoxic activities of Co (II), Ni (II) and Cu (II) complexes with 1, 2, 4-triazole Schiff bases. European Journal of Medicinal Chemistry. 2008 Dec 1;**43**(12):2639-2649

[14] Andiappan K, Sanmugam A, Deivanagam E, Karuppasamy K, Kim H-K, Vikraman. *In vitro* cytotoxic activity of novel Schiff base ligand-lanthanide complexes. Nature Scientific Reports. 2018;**8**. DOI: 10.1038/s41598-018-21366-1

[15] Murtaza S, Akhtar MS, Kanwal F, Abbas A, Ashiq A, Shamim S. Synthesis and biological evaluation of schiff bases of 4-aminophenazone as an anti-inflammatory, analgesic and antipyretic agent. Journal of Saudi Chemical Society. 2017;**21**:S359-S-372

[16] Ohn N, Kim JG. Mechanochemical Post-Polymerization Modification: Solvent-Free Solid-State Synthesis of Functional Polymers. ACS Macro Letters. 2018;7:561-565

[17] Ottolenghi M, Sheves M. Molecular Mechanism for the Function of Bacteriorhodopsin and Visual Pigments: Studies with Model Compounds and Artificial Pigments. In: Chemistry and Biology of Synthetic Retinoids. CRC Press; 2017 Dec 22. pp. 99-123

[18] Singh J, Kaur V, Singh R, Bhardwaj VK. Exploration of solvent responsive Cr^{3+}-Schiff base conjugates for monitoring Cr^{3+} ions and organophosphates: Fabrication of spot-testing devices. Spectrochimica Acta Part A: Molecular and Biomolecular Spectroscopy. DOI: 10.1016/j.saa.2018.04.056

[19] Hasani B, Zamani A, Moftakhar MK, Mostafavi M, Yaftian MR, Ghorbanloo M. Ionophore Properties of Schiff Base Compounds as Ion Sensing Molecules for Fabricating Cu (II) Ion-Selective Electrodes. Journal of Analytical Chemistry. 2018 Jan;**73** (1, 1):82-90

[20] Hoffmann RC, Sanctis S, Schneider JJ. Molecular Precursors for ZnO Nanoparticles: Field-Assisted Synthesis, Electrophoretic Deposition, and Field-Effect Transistor Device Performance. Inorganic Chemistry. 2017 Jun 9;**56**(13):7550-7557

[21] Carreño A, Zuñiga C, Paez-Hernandez D, Gacitua M, Polanco R, Otero C, et al. Study of structure-bioactivity relationship of three new pyridine Schiff bases: Synthesis, spectral characterization, DFT calculations and biological assays. New Journal of Chemistry. 2018

[22] Wang X, Fan R, Dong Y, Su T, Huang J, Du X, et al. Metal (II)-Induced Synthesis of Asymmetric Fluorescence Benzimidazoles Complexes and Their Dye-Sensitized Solar Cell Performance as Cosensitizers. Crystal Growth & Design. 2017 Aug 30;**17**(10):5406-5421

[23] Kumbar SS, Hosamani KM, Gouripur GC, Joshi SD. Functionalization of 3-chloroformylcoumarin to coumarin Schiff bases using reusable catalyst: an approach to molecular docking and biological studies. Royal Society Open Science. 2018 May 1;**5**(5):172416

[24] Chen D, Tang Q, Zou J, Yang X, Huang W, Zhang Q, Shao J, Dong X. pH-Responsive PEG–Doxorubicin-Encapsulated Aza-BODIPY Nanotheranostic Agent for Imaging-Guided Synergistic Cancer Therapy. Advanced healthcare materials; 2018 Jan 15

[25] Malinowski M, Iwan A, Tazbir I, Boharewicz B, Sikora A, Stafiniak A. Polyazomethines and their acid–base interactions with Nafion and Nafion–imidazole membranes for efficient fuel cells. Sustainable Energy & Fuels. 2017;**1**(8):1810-1819

[26] Roy S, Mondal TK, Layek A, Saha R, Sinha C. Structure, spectra and electrical conductivity of copper (I) and silver (I) phosphino bridging mixed ligand complexes with coumarinyl Schiff base. Inorganica Chimica Acta. 2018 Jan 1;**469**:523-535

[27] Niakan M, Asadi Z, Masteri-Farahani M. A covalently anchored Pd (II)-Schiff base complex over a modified surface of mesoporous silica SBA-16: an efficient and reusable catalyst for the Heck-Mizoroki coupling

reaction in water. Colloids and Surfaces A: Physicochemical and Engineering Aspects. 2018 May 1

[28] Pratihar JL, Mandal P, Brandão P, Mal D, Felix V. Synthesis, characterization, spectral and catalytic activity of tetradentate (NNNO) azo-imine Schiff base copper (II) complexes. Inorganica Chimica Acta. 2018 Apr 23

[29] Barwiolek M, Sawicka J, Babinska M, Wojtczak A, Surdykowski A, Szczesny R, et al. Synthesis, spectroscopic characterization, fluorescence properties of new silver (I) complexes with optically active Schiff bases and their application in thin layers deposition. Polyhedron. 2017 Oct 15;**135**:153-160

[30] Liu G, Cogan DA, Owens TD, Tang TP, Ellman JA. Synthesis of enantiomerically Pure N-tert-butanesulfinyl imines (tert-butanesulfinimines) by the direct condensation of tert-butanesulfinamide with aldehydes and ketones. The Journal of Organic Chemistry. 1999;**64**:1278-1284

[31] Campaigne EI. Thiophenes and their benzo derivatives: (iii) synthesis and applications. Netherlands: Elsevier BV; 1984

[32] Andrade CKZ, Takada SCS, Alves LM, Rodrigues JP, Suarez PAZ, Brandão RF, et al. Molecular sieves in ionic liquids as an efficient and recyclable medium for the synthesis of imines. Synlett. 2004;(12):2135-2138

[33] Huang B, Tian H, Lin S, Xie M, Yu X, Xu Q. Cu(I)/TEMPO-catalyzed aerobic oxidative synthesis of imines directly from primary and secondary amines under ambient and neat conditions. Tetrahedron Letters. 2013;**54**:2861-2864

[34] Houben J, Fischer W. Formation of aromatic nitriles by basic hydrolysis of trichloromethyl aryl ketimines. Acidic hydrolysis yields ketones. Journal für Praktische Chemie. 1929;**123**:262-313

[35] Pickard PL, Tolbert TL. An improved method of ketimine synthesis. The Journal of Organic Chemistry. 1961;**26**:4886-4888

[36] Mosher H, Blanz JE. Notes—Reduction of o-bromoanisole by lithium dineopentylamide. The Journal of Organic Chemistry. 1957;**22**:445-446

[37] Langheld K. Über das Verhalten von α-Aminosäuren gegen Natriumhypochlorit (in German). Berichte der Deutschen Chemischen Gesellschaft. 1909;**42**:2360-2374

[38] Barluenga J, Aznar F, Valdes C. N-trialkylsilylimines as coupling partners for Pd-catalyzed C-N bond-forming reactions: One-step synthesis of imines and azadienes from aryl and alkenyl bromides. Angewandte Chemie, International Edition. 2004;**43**:343-345

[39] Zhu W, Mena M, Jnoff E, Sun N, Pasau P, Ghosez L. Multicomponent reactions for the synthesis of complex piperidine scaffolds. Angewandte Chemie, International Edition. 2009;**48**:5880-5883

[40] Mekheimer RA, Hameed AM, Sadek KU. First synthesis and isolation of the E-and Z-isomers of some new Schiff bases. Reactions of 6-azido-5-formyl-2-pyridone with aromatic amines. Molecules. 2008 Jan 30;**13**(1):195-203

[41] Smith S, Bou-Abdallah F. The Kinetics of the Cis-to-Trans Thermal Isomerization of 4-Anilino-4'-Nitroazobenzene are Highly Influenced by Solvent Polarity. Journal of Thermodynamics & Catalysis. 2017;**8**:181-186

[42] Joshi NK, Fuyuki M, Wada A. Polarity controlled reaction path and kinetics of thermal cis-to-trans

isomerization of 4-aminoazobenzene. The Journal of Physical Chemistry B. 2014 Feb 5;**118**(7):1891-1899

[43] Wazzan NA, Richardson PR, Jones AC. Cis-trans isomerisation of azobenzenes studied by laser-coupled NMR spectroscopy and DFT calculations. Photochemical & Photobiological Sciences. 2010;**9**(7):968-974

[44] Angelini G, Canilho N, Emo M, Kingsley M, Gasbarri C. Role of solvent and effect of substituent on azobenzene isomerization by using room-temperature ionic liquids as reaction media. The Journal of Organic Chemistry. 2015 Jul 9;**80**(15):7430-7434

[45] Kaiser EW, Pierce DS. Study of the Thermodynamics (Thermal and Cl Catalyzed) and Kinetics of the Cis and Trans Isomerizations of CF_3CF= CHF, CF_3CH=$CHCF_3$, and CH_3CH=$CHCH_3$ in 100-950 Torr of N_2 Diluent at 296-875 K: Effect of F and CF_3 Substitution on the Isomerization Process Including the Fluorine "Cis-Effect". The Journal of Physical Chemistry A. 2015 Aug 12;**119**(34):9000-9017

[46] Rodríguez MR, Del Plá J, Piro OE, Echeverría GA, Espino G, Pis-Diez R, et al. Structure, tautomerism, spectroscopic and DFT study of o-vanillin derived Schiff bases containing thiophene ring. Journal of Molecular Structure. 2018 Aug 5;**1165**:381-390

[47] Brkić DR, Božić AR, Marinković AD, Milčić MK, Prlainović NŽ, Assaleh FH, et al. Detailed solvent, structural, quantum chemical study and antimicrobial activity of isatin Schiff base. Spectrochimica Acta Part A: Molecular and Biomolecular Spectroscopy. 2018 May 5;**196**:16-30

[48] Sobczyk L, Chudoba D, Tolstoy PM, Filarowski A. Some brief notes on theoretical and experimental investigations of intramolecular hydrogen bonding. Molecules. 2016 Dec 2;**21**(12):1657

[49] Liu X, Manzur C, Novoa N, Celedón S, Carrillo D, Hamon JR. Multidentate unsymmetrically-substituted Schiff bases and their metal complexes: Synthesis, functional materials properties, and applications to catalysis. Coordination Chemistry Reviews. 2018 Feb 15;**357**:144-172

[50] Gupta KC, Sutar AK, Lin CC. Polymer-supported Schiff base complexes in oxidation reactions. Coordination Chemistry Reviews. 2009 Jul 1;**253**(13-14):1926-1946

[51] Leng Y, Liu J, Zhang C, Jiang P. A polyhedral oligomeric silsesquioxane (POSS)-bridged oxo-molybdenum Schiff base complex with enhanced heterogeneous catalytic activity in epoxidation. Catalysis Science & Technology. 2014;**4**(4):997-1004

[52] Brkić DR, Božić AR, Marinković AD, Milčić MK, Prlainović NŽ, Assaleh FH, et al. Detailed solvent, structural, quantum chemical study and antimicrobial activity of isatin Schiff base. Spectrochimica Acta Part A: Molecular and Biomolecular Spectroscopy. 2018 May 5;**196**:16-30

[53] Gaballa AS, Asker MS, Barakat AS, Teleb SM. Synthesis, characterization and biological activity of some platinum (II) complexes with Schiff bases derived from salicylaldehyde, 2-furaldehyde and phenylenediamine. Spectrochimica Acta Part A: Molecular and Biomolecular Spectroscopy. 2007 May;**67**(1, 1): 114-121

[54] Abu-Dief AM, Mohamed IM. A review on versatile applications of transition metal complexes incorporating Schiff basesBeni-Suef University Journal of Basic and Applied Sciences. 2015 Jun 1;**4**(2):119-133

[55] Nasr-Esfahani M, Zendehdel M, Nia NY, Jafari B, Babadi MK. Fabrication and characterization of a new dye sensitized solar cell with a new Schiff base cobalt complex as a redox mediator. RSC Advances. 2014;**4**(31):15961-15967

[56] Hameed A, al-Rashida M, Uroos M, Abid Ali S, Khan KM. Schiff Bases in medicinal chemistry: a patent review (2010-2015). Expert Opinion on Therapeutic Patents. 2017 Jan 2;**27**(1):63-79

[57] Prashanthi Y, Kiranmai K, Subhashini NJ. Synthesis, potentiometric and antimicrobial studies on metal complexes of isoxazole Schiff bases. Spectrochimica Acta Part A: Molecular and Biomolecular Spectroscopy. 2008 Jun 1;**70**(1):30-35

[58] Ebrahimi HP, Hadi JS, Almayah AA, Bolandnazar Z, Swadi AG, Ebrahimi A. Metal-based biologically active azoles and β-lactams derived from sulfa drugs. Bioorganic & Medicinal Chemistry. 2016 Mar 1;**24**(5):1121-1131

[59] Shebl M. Synthesis, spectral studies, and antimicrobial activity of binary and ternary Cu(II), Ni(II), and Fe(III) complexes of new hexadentate Schiff bases derived from 4, 6-diacetylresorcinol and amino acids. Journal of Coordination Chemistry. 2009 Oct 10;**62**(19):3217-3231

[60] Guo Z, Xing R, Liu S, Yu H, Wang P, Li C, et al. The synthesis and antioxidant activity of the Schiff bases of chitosan and carboxymethyl chitosan. Bioorganic & Medicinal Chemistry Letters. 2005 Oct 15;**15**(20):4600-4603

[61] Bandgar BP, Gawande SS, Bodade RG, Totre JV, Khobragade CN. Synthesis and biological evaluation of simple methoxylated chalcones as anticancer, anti-inflammatory and antioxidant agents. Bioorganic & Medicinal Chemistry. 2010 Feb 1;**18**(3):1364-1370

[62] Zhang N, Fan YH, Zhang Z, Zuo J, Zhang PF, Wang Q, et al. Syntheses, crystal structures and anticancer activities of three novel transition metal complexes with Schiff base derived from 2-acetylpyridine and l-tryptophan. Inorganic Chemistry Communications. 2012 Aug 31;**22**:68-72

[63] Li LJ, Wang C, Tian C, Yang XY, Hua XX, Du JL. Water-soluble platinum (II) complexes of reduced amino acid Schiff bases: synthesis, characterization, and antitumor activity. Research on Chemical Intermediates. 2013 Feb 1; **39**(2):733-746

[64] Sriram D, Yogeeswari P, Myneedu NS, Saraswat V. Abacavir prodrugs: microwave-assisted synthesis and their evaluation of anti-HIV activities. Bioorganic & Medicinal Chemistry Letters. 2006;**16**(8):2127-2129

[65] Alam MS, Choi JH, Lee DU. Synthesis of novel Schiff base analogues of 4-amino-1,5-dimethyl-2-phenylpyrazol-3-one and their evaluation for antioxidant and anti-inflammatory activity. Bioorganic & Medicinal Chemistry. 2012;**20**:4103

[66] Arabahmadi R, Orojloo M, Amani S. Azo Schiff bases as colorimetric and fluorescent sensors for recognition of F^-, Cd^{2+} and Hg^{2+} ions. Analytical Methods. 2014;**6**(18):7384-7393

[67] Al Zoubi W, Al Mohanna N. Membrane sensors based on Schiff bases as chelating ionophores–A review. Spectrochimica Acta Part A: Molecular and Biomolecular Spectroscopy. 2014 Nov 11;**132**:854-870

[68] Zhang J, Xu L, Wong WY. Energy materials based on metal Schiff base complexes. Coordination Chemistry Reviews. 2017 Sep 6. DOI: 10.1016/j.ccr.2017.08.007

Chapter 7

Salen and Related Ligands

Ashish K. Asatkar, Mamta Tripathi and Deepali Asatkar

Abstract

The salen and related ligands are very popular among the inorganic chemists due to multiple reasons such as ease in synthesis, coordinating ability with very long range of metal ions, facilitating the metal ions to adopt various geometries, ability of stabilising the metal ion in variable oxidation states and potential applications of metallosalen in several fields. The most common application of metallosalen is in the field of catalysis because of their recoverability, reusability, high efficiency, high selectivity and their capability of working as homogeneous as well as heterogeneous catalysts for numerous functional group manipulations including asymmetric synthesis. Molecular magnetism, sensory applications, bioinorganic activities and medicinal applications of metallosalen are also very promising areas of their applications. Porous materials involving metal organic frameworks (MOFs) and supramolecular building blocks are increasingly getting attention of researchers for the gas absorption and heterogeneous catalysis.

Keywords: salen, salphen, Schiff-base, chelate ligand, metallosalen

1. Introduction

The coupling of aldehyde group with primary amine yields imine bond which is called Schiff's base. Salen ligand system, one of the most studied classes of chelate ligands, is also a Schiff's base ligand. The earliest report of salen-metal complexes is probably by Pfeiffer et al. in the year 1933 [1]. The word 'salen' is composed of two abbreviations, sal+en; '*sal*' stands for salicylaldehyde and '*en*' stands for ethylenediamine. When two equivalents of salicylaldehyde reacts with one equivalent of ethylenediamine potential tetradentate chelating ligand known as 'salen' is produced (**Figure 1**).

Usually, these reactions do not need any catalyst and proceed straightforwardly but sometimes the products may be hydrolysed in reversible manner. To overcome this problem, dehydrating agents or molecular sieves (3 Å) are used so that the water molecules produced during the reaction can be absorbed. Dean Stark apparatus is also used for the removal of water molecules when water-immiscible solvent (e.g., toluene or benzene) is used. Sometimes template synthesis is also performed to get metal-salen complexes directly in which process first metal-salicylaldehyde complex is prepared *in-situ* as template then ethylenediamine is added to get salen ligand. Although, the salen ligands are sensitive towards hydrolysis which is catalysed by acid, their metal complexes are quite stable and thus to avoid the hydrolysis of salen ligands during the applications, their metal-complexes are often used. Metal salen can work even in aqueous medium. Moreover, the salen ligands have potential to stabilise metal ions in various oxidation states, making them good candidates as catalysts.

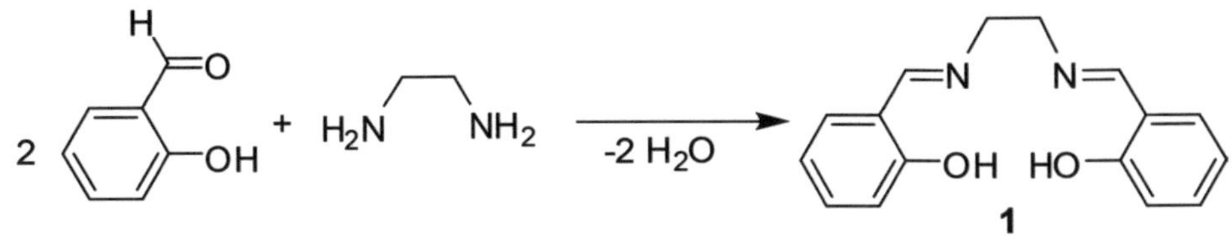

Figure 1.
Synthesis of salen ligand.

Salen ligand possess N_2O_2 donor sites which offers metal ions to adopt various geometries such as square planar, tetrahedral, square pyramidal and octahedral as well, with additional ligand(s) if required. A large number of metal ions have been introduced to salen to produce variety of complexes [2–4]. A very broad range of transition metals, main group metals and inner transition metals have been coordinated with salen ligand systems. Being the multidentate ligand, their complexes often have very high formation constants. Salen based complexes have potentially been used in several fields like catalysis, biochemistry, electrochemistry, sensors, molecular magnetism and materials science. Salen-metal complexes are still leading in the field of homogeneous catalysis for various organic reactions. In the past few decades, numerous reviews based on salen ligand system have been published, highlighting its importance [5–8].

2. Salen ligands and derivatives

Several manipulations have been done on parent salen system to develop the varieties of salen system for various applications. The derivatives of salen are designed to develop desirable properties like solubility, stability, chirality, catalysis, extended conjugation, etc. Aromatic ring and diamine linkage (e.g., ethylene link) are two main portions in salen ligand system, which are used to put various substituents. 3-,5-Positions of salicylideneimine are frequently used for substitution. Substitution at 3- and 5-positions of salicylideneimine also improves the catalytic activities and prevents dimerization as well. The numbering of positions in salen system is shown in **Figure 2**. Substitution at aromatic ring of salicylaldehyde is very popular to enhance solubility of salen ligand and its metal complexes while the substitution at diamine linkage is commonly used to get the chiral ligand. Another position available for the substitution is carbon atom of imine bond.

2.1 Chiral salen

The asymmetry is introduced to salen system mostly by the use of chiral diamine. Chiral salen are of particular importance in asymmetric synthesis as enantioselective catalyst. Many procedures are known for chiral synthesis of

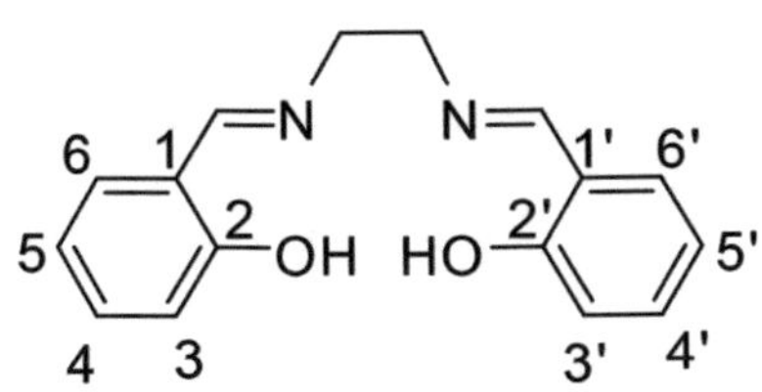

Figure 2.
Numbered positions in salen ligand.

ligands using diamine having one or more stereocentres [9, 10], or a stereoaxis [11], through the incorporation of axial [12] or planar [13–15] chirality within the salicylaldehyde. *Trans*-1,2-diaminocyclohexane and 1,2-diphenylethylene-1,2-diamine are often used as 1,2-diamine to produce the chiral salen. These two chiral salen (**2** and **3**) are very popular and their several derivatives have been reported [16]. Very often, tertiary butyl group and long alkyl chain are put to modify solubility, steric factor and electronic factor.

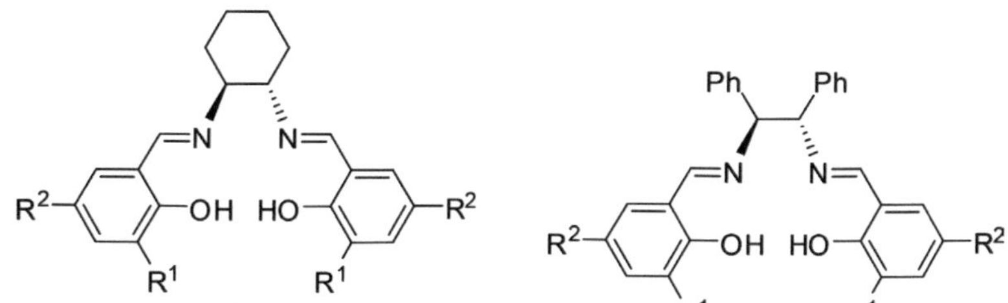

2 **3**

Chiral binaphthyl salen complexes (**4** and **5**) have been designed in such a way that the complexes possess two stereogenic centres and thus considered as second generation metal salen complexes. One of the stereogenic centres belongs to binaphthyl unit while other belongs to diamine unit [17–19]. The complexes were used for non-racemic oxidation of prochiral sulphides.

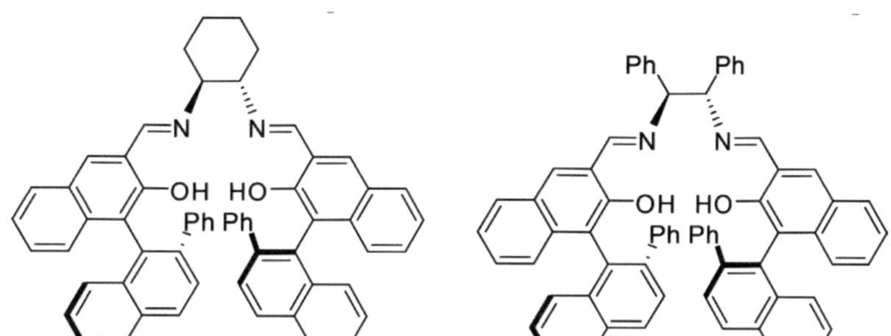

4 **5**

2.1.1 Non-symmetrical salen

Salen ligand systems have successfully been employed as homogeneous catalysts for variety of organic functional group manipulations. Very often they are symmetrical and having C_2-axis of symmetry. Non-symmetrical ligands bring out further magnify opportunities for tuning of electronic, steric and catalytic properties and therefore various nonsymmetrical analogues of salen have also been developed [20]. There are various advantages of unsymmetrical salen over symmetrical salen such as nonsymmetrical salen with single functional group can be immobilised onto heterogeneous and homogeneous traps to recover it after use [21, 22]. Moreover, electron releasing and/or withdrawing groups can be put on aryl rings of salicylideneimine part of salen. Presence of electron releasing and withdrawing groups together acts as push-pull system for electron density. Also, the unsymmetrical salen-metal complexes have shown better enantioselectivity in certain cases [23, 24].

The easiest way to prepare an unsymmetrical salen can be direct two step Schiff base coupling i.e., the reaction between salicylaldehyde and ethylenediamine in 1:1 molar ratio to get mono-keto-imine product followed by the reaction with substituted salicylaldehyde (**Figure 3**) [25–27]. This method do not need any protection of group or presence of special reagent, but the main drawback of this method is that the stepwise coupling is not much favourable due to the formation of symmetrical product in first step and lability of imine bonds towards hydrolysis which reduces the yield of desirable unsymmetrical product drastically. Jacobsen et al. exhibited another way to prepare nonsymmetrical salen ligand directly by the reaction of two different salicylaldehyde derivatives and (1R,2R)-(+)-1,2-diaminocyclohexane L-tartrate in 1:1:1 molar ratio in single spot, but in moderate yield (**Figure 4**) [28, 29]. Another approach for the synthesis of non-symmetrical salen is selective protection of one of the amine groups of diamine compound followed by Schiff base coupling of another amine group with salicylaldehyde, then the protected amine group is deprotected and coupled with distinct salicylaldehyde (**Figure 5**) [30, 31].

Silica- and polymer-immobilised Co(III)-salen non-symmetrical complexes (**6**) have also been developed and successfully used as catalysts for hydrolytic kinetic resolution of terminal epoxides with better rate, enantioselectivity and recyclability [32, 33]. Similar Mn(III)-salen non-symmetrical complexes have also been designed and studied [7]. Rigamonti et al. reported the synthesis of nonsymmetrical salen-Cu(II) complexes (7–**14**) by the reaction of salicylaldehyde/5-nitrosalicylaldehyde

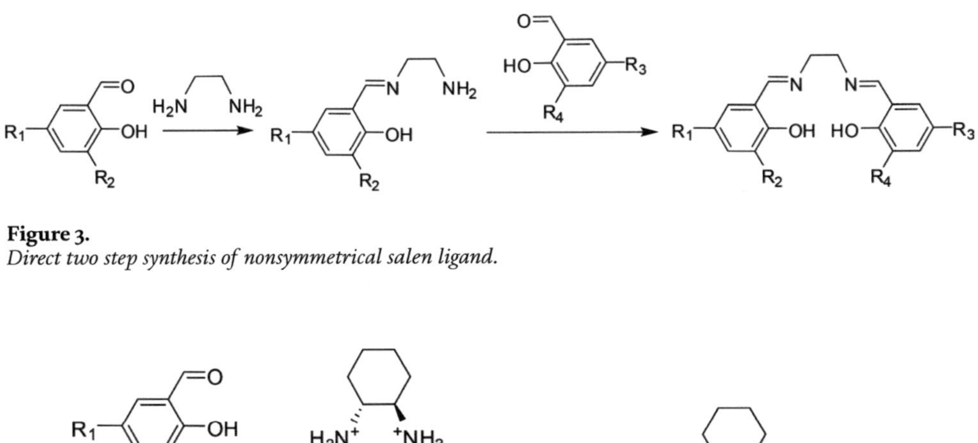

Figure 3.
Direct two step synthesis of nonsymmetrical salen ligand.

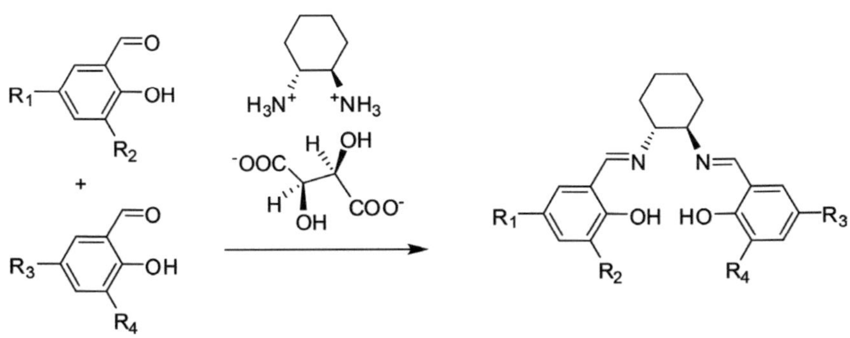

Figure 4.
Direct one step synthesis of nonsymmetrical salen ligand.

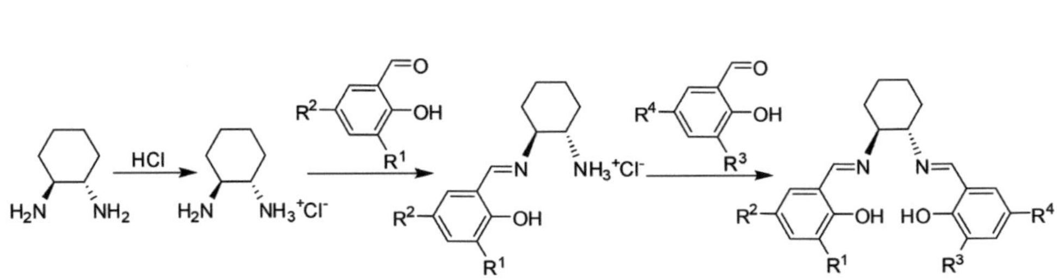

Figure 5.
Protection-deprotection method for the synthesis of nonsymmetrical salen ligand.

and ethylenediamine/propylenediamine in 1:1 molar ratio in presence of Cu(II) ion and pyridine followed the addition of differently substituted salicylaldehyde and their nonlinear optical properties were studied and correlated with the structural diversities [34]. Salen ligand with methyl group at ethylene backbone is known as "salpn" (**15**). Salpn and its complexes have been used as additive in engine oil [35].

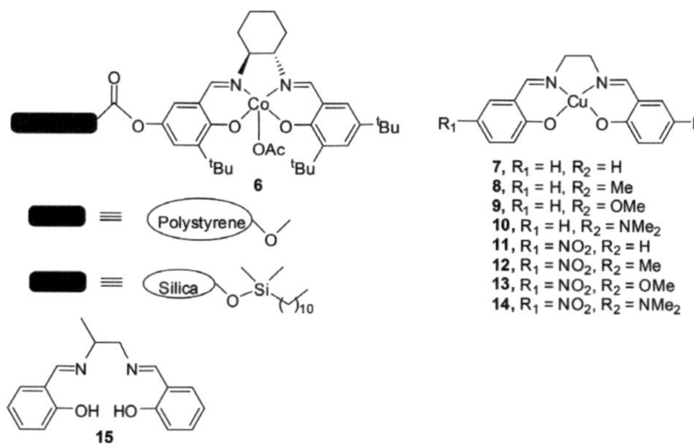

2.2 Conjugated salen

When phenylenediamine (phen) is taken in place of ethylenediamine during the reaction, the ligand formed is known as "Salphen" or sometimes "Salophen" (**16**). Salphen has extended conjugation with rigid planarity when coordinated with metal ion in square planar, octahedral or square pyramidal geometry, which is a very important criterion for material applications. Their photophysical properties can be fine-tuned by putting suitable substituents. Pietrangelo et al. synthesised thiophene capped salen ligands and their V, Ni and Cu copper complexes (**17**) and electrochemically polymerised them [36]. Asatkar et al. reported the synthesis of thiophene analogue of salphen (**18**) by taking 2-formyl-3-hydroxythiophene in place of salicylaldehyde and their Cu(II) and Zn(II) complexes [37]. However, the complexes could not be electrochemically polymerised as thiophene capped salphen did.

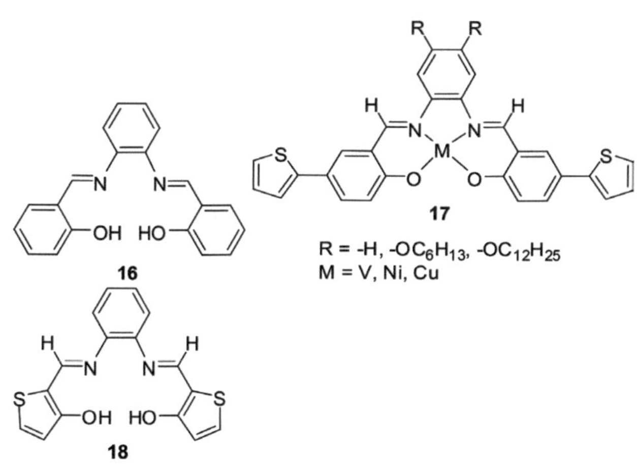

Even more complicated salphen have been developed by linking/merging two or more such units either through phenelene or salicylaldehyde [38] Bis-salphen scaffold ligand can be prepared by the reaction of four equivalents of salicylaldehyde and one equivalent of 1,2,4,5-benzenetetramine and its derivatives can also be developed is similar way [39, 40]. Kleij et al. reported the synthesis of unsymmetrical bis-metal-salphen scaffold complexes by partial hydrolysis of parent symmetrical bis-zinc-salphen scaffold complex followed by Schiff-base coupling with differently substituted salicylaldehyde derivatives (**19–29**) [41]. Similarly, another bis-salphen symmetrical and unsymmetrical ligands (**30**) are prepared using one equivalent of 3,3′-diaminobenzidene and four equivalents of salicylaldehyde [42, 43]. Salphen based tri [3+3] (**31**), tetra [4+4] and hexa [6+6] macrocycles have also been prepared using 2,3-dihydroxybenzene-1,4-dicarbaldehyde and 1,2-phenylenediamine [44–47].

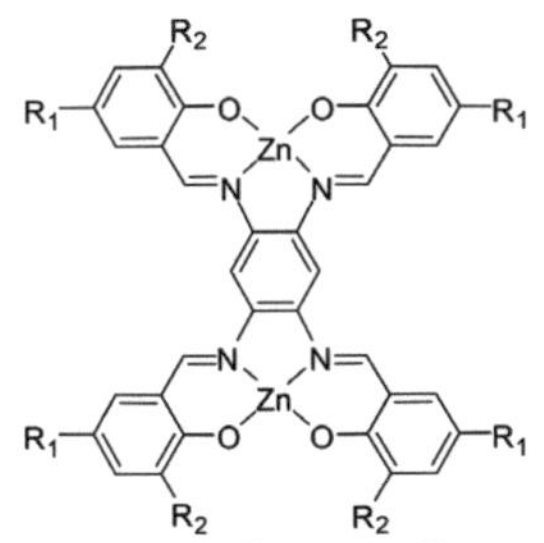

19, $R_1 = R_2 = H$
20, $R_1 = NO_2$, $R_2 = H$
21, $R_1 = NO_2$, $R_2 = NO_2$
22, $R_1 = Br$, $R_2 = H$
23, $R_1 = CH_3$, $R_2 = H$
24, R_1 = allyl, $R_2 = H$
25, $R_1 = OCH_3$, R_2 = allyl
26, $R_1 = H$, $R_2 = Br$
27, $R_1 = OC_2H_5$, $R_2 = H$
28, $R_1 = CF_3$, $R_2 = H$
29, $R_1 = Cl$, $R_2 = Cl$

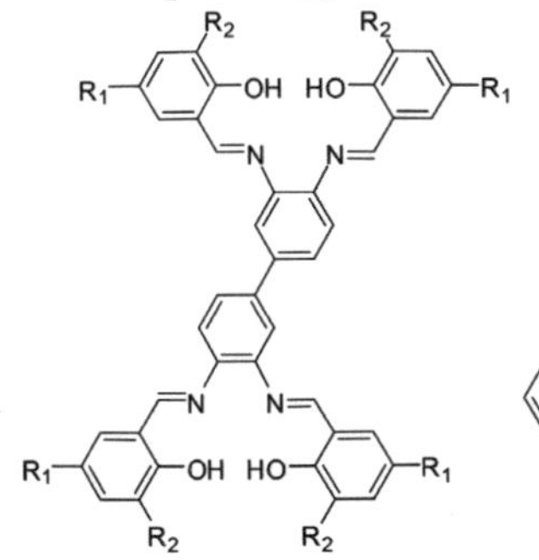

30

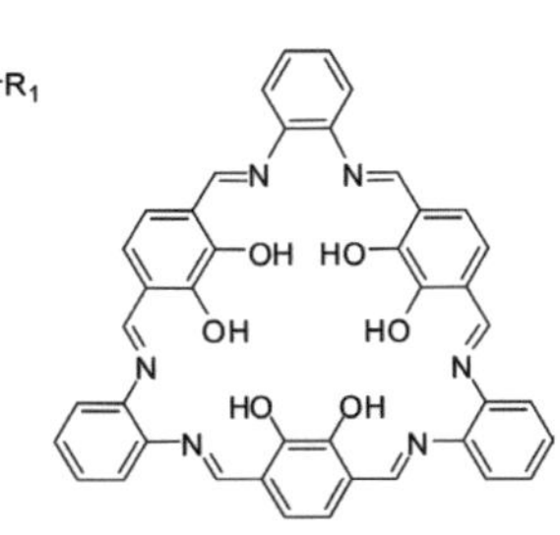

Tri-salphen
[3+3]
31

2.3 Salen based metal organic framework

Metal-organic frameworks (MOFs), is a fascinating classification of porous materials that can exits as self-assembled *via* coordination of metal aggregation/ions with organic linkers [48–50]. Shultz et al. synthesised MOF using pyridine functionalized Salen-Mn complex and tetrakis(4-carboxyphenyl)benzene [51]. The MOF was further used to prepare new MOFs with change in metal ion. The Mn-MOF was demetalated first using H_2O_2 then remetalated with Cr(II), Co(II), Ni(II), Cu(II) and Zn(II) ions [52]. Lin et al. reported MOFs using chiral Mn-Salen functionalized with variable size dicarboxalic acid linkage. The MOFs exhibited asymmetric epoxidation catalysis with enantiomeric excess as high as 92% [53]. Jeon et al. reported infinite coordination particles based on carboxalic acid functionalized Salen-Zn complex and studied the gas absorption capacity. The amorphous material showed excellent hydrogen gas intake capability [54]. Roesky et al. used carboxalic acid functionalized Salen-Ni complex and lanthanides to synthesise MOFs [55]. Shape of the framework was found to be dependent of size of lanthanides.

Kleij et al. found the unique self-aggregation nature of bis-Zn(salophen) [14, 15, 56, 57]. They have secure self-assembly behaviour through linking coordination motifs that are fundamentally different from those usually found for the self-assembly of mononuclear Zn-salophens [58]. This takes place on both at the interface of solid-liquid as well in solution. Oligomeric $(Zn—O)_n$ coordination moiety are accustomed inside the assembly and this is quite distinct from mononuclear analogues

DOI: http://dx.doi.org/10.5772/intechopen.88593

of Zn(salphen) which form dimeric structures having a classical Zn_2O_2 central unit [59]. Multimetallic salen frameworks have been revealed to act as metallohosts forming adduct complexes with further structural ordering upon substrate binding [38]. Nabeshima et al. employed a linear metallohost containing two N_2O_2 binding units [60]. Upon metalation with Zn(II) a 1:3 ligand to metal complex forms *via* a highly cooperative process. One Zn(II) ion is situated in a C-shaped O_6 site in the centre of the helical complex. Guest exchange was shown to occur through substitution of the central Zn(II) with rare earth metal and lanthanide cations. Excitingly, the helicity of the complex is relying on the size of the central guest cation.

3. Analogues of salen

Due to the extended applications of salen ligand systems, their various analogues have been developed and studied. Chalcogen analogues of salen include sulphur and selenium derivatives as thiasalen and selenasalen. However, the sulphur and selenium analogues are relatively less explored because of the volatile nature, instability, synthetic complications, unpleasant smell and adverse effect of thiol and selenol compounds. To synthesise the metal-thiasalen/selenasalen complexes, template synthesis is often used.

Dutta et al. reported the one pot synthesis of thia/selena analogues of salen-metal complexes (**32–37**) *via* oxidative addition of zero valent group ten metals (Ni(0), Pd(0) and Pd(0)) to S-S/Se-Se bond of bis(o-formylphenyl)disulphide/–diselenide followed by *in situ* coupling with ethylenediamine [61]. Panda et al. reported the synthesis of bis(alkylseleno)salen ligands (**38–41**) by the reaction of 2-(alkylthio/seleno)benzaldehyde and ethylenediamine [62]. Their complexation with Pd(II) and Pt(II) ions exhibited very interesting results. Complexation of 2-(alkylseleno) benzaldehyde with Pd(II) and Pt(II) ion yielded the formation of unsymmetrical complexes with the cleavage of one of the alkyl groups from Se-C(alkyl) bonds. However, the complexation with Pd(II) ions Complexation of 2-(methylthio) benzaldehyde with Pt(II) ion, reported by Dutta et al., yielded similar unsymmetrical complex (**42–46**) while the same with Pd(II) ion yielded time dependent product [63]. When the reaction mixture was refluxed for 5 min the symmetrical complex (**48**) with both the methyl groups intact was obtained, but when it was refluxed for 4 h the unsymmetrical complex (**47**) was obtained.

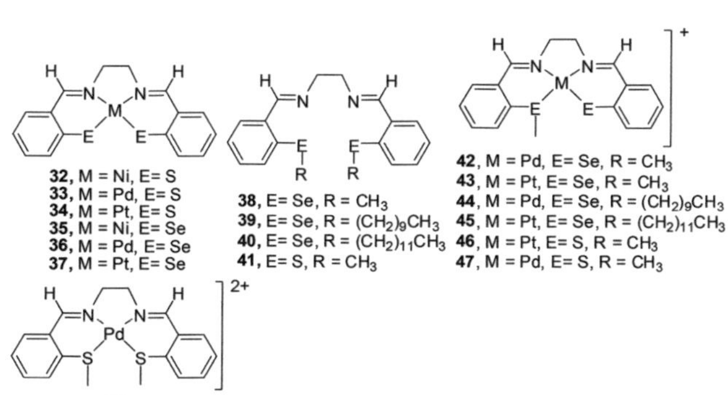

Benzene rings have also been replaced by other aromatic rings to design the new salen analogues. Jeong et al. reported the synthesis of pyridine based salen type chiral ligands (**49–50**) and their complexes and used them as enantioselective catalysts in Henry reaction [64]. Asatkar et al. reported the thiophene analogues (**51–52**)

of salen ligand system [65]. Interestingly, thiophene analogue of simple salen was found to exist in different tautomeric forms in solid and solution phases, unlike salen ligand. Its reaction with Cu(II) ion resulted in the dimeric complex. Another example of change in aromatic ring is pyrrole based salen type ligand (**53**), reported by Berube et al. along with its dimeric samarium(II) complex [66].

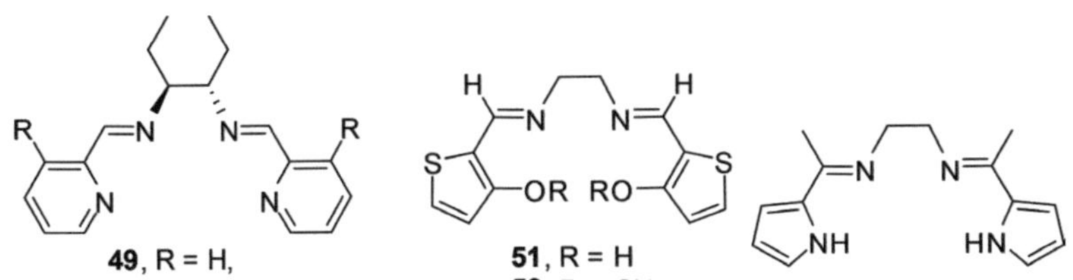

49, R = H,
50, R = CH_3

51, R = H
52, R = CH_3

53

4. Applications of salen-metal complexes

M(salen) complexes have unique and exciting class of ligand based complexes with exceptionally versatile applications ranging from laboratory reaction to mass scale industries level. Interestingly, metal salen complexes gained popularity because of their roles in multiple areas few important of them are discussed below:

4.1 Catalysis

Metal-salen complexes appear as both homogeneous and heterogeneous catalyst and have been substantially investigated by researchers for multiple uses [5]. The most attracting feature of metal salen catalysts is that they can be recovered and reused. Usually found that the salen as catalyst possess high stability revealed by their high stability constants [7]. When metal salen are applied as catalyst, demetalation of the complex occurs because of competitive binding with reagents, solvent or products, may be associated with changes in the oxidation state of metal in catalytic cycle. Few important reactions catalysed by metal salen includes Meerwein-Ponndorf-Verley reductions (MPV) [67, 68], Friedel-Crafts Reactions [69], Oppenauer oxidation, Tishchenko reactions [70, 68], ene reaction [71], mixed-aldol condensation [72, 73], Diels-Alder reactions [71], dipolar cycloadditions, Claisen rearrangements [74] and the cyclotrimerization of isocyanates to isocyanurates [75].

Interestingly, Metal salen holds important role in many oxidation reactions like alkene epoxidation [76], asymmetric syntheses of cyanohydrins and amino acids [77], and oxidation of heteroatom-containing compounds [78]. In biological system they actively take part in catalytic oxidation of Ni(III) oxidised in the catalytic cycles of Ni-Fe hydrogenases [79–82], acetyl coenzyme A synthase(ACS) [83–85], COdehydrogenase [86, 87], and methyl coenzyme M reductase [88]. Mirkhani et al. have found that the oxidation of diphenyl sulphide mediated by Mn(III)-salphen and Mn(III)-salen employing terminal oxidant as sodium periodate. The Mn(III)-salphen complex yields a product mixture of sulfoxide and sulfone (4, 1 ratio) in 100% transformation under mild conditions [89]. This is in contrast to the analogous Mn(III)salen complex which only led 18% (ratio of sulfoxide and sulfone, 2:1). Mn(III)-salphen catalytic system was also successfully applied towards a variety of other sulphides and also furnished 100% yields.

Salen complex of heterobimetallic origin have been exclusively examined for many asymmetric catalytic synthesis [90]. Salen ligands are prepared from diamines and salicylaldehydes [91], configuration of both of these constituents can easily be changed, sterically modified as per desirable physical and electronically altered which makes it possible for the synthesis of recyclable and immobilised salen complexes [7, 92–96]. Shibasaki et al. have used chemoselective complexation of transition metals at N_2O_2 coordination core while the rare earth metal utilised O_2O_2 core of same ligand. However, the key role for selectivity and reactivity of these multimetallic catalysts is based on metal ions e.g., coupling of Cu(II) and Sm(III) yields 66–99% enantiomeric excess (ee) in Mannich-type reactions [97] whereas Pd(II) and La(III) is the best combination for the asymmetric synthesis in Henry reaction, yielding product in 72–92% ee [98].

4.2 Molecular magnetism

Magnetic linkage of paramagnetic metal centres with some non-innocent ligands, in multimetallic salen complexes has produced essential information on spin interaction mechanisms. The extent of magnetic interaction (whether it be antiferromagnetic or ferromagnetic) is dependent on a number of factors including the distance between the paramagnetic centres and comparative orientation of the related magnetic orbitals. The relative ease of synthesis and the distance between the paramagnetic centres. Single molecule magnets have gained much research attention since the discovery of spontaneous magnetization below a critical temperature [99, 100]. By applying proper ligand scaffolds, ferromagnetic interactions can be enforced between metal centres in multimetallic complexes [101]. Glaser et al. investigated phloroglucinol as a linker between paramagnetic metal salen units [102–104]. At the time, *m*-phenylene linkers had been well established in the organic radical community as an efficient ferromagnetic coupler and had been used extensively as a means to produce high spin organic radicals [105]. First row of transition metal V(IV)=O [106], Mn(III) [107], Fe(III) [108], Ni(II) [109] and Cu(II) [110] are best fitted coordinating with triple salen.

4.3 Material applications

Metal salen based materials have drawn attraction of material scientists as well [111]. Metal organic framework (MOF) and zeolite encapsulated salen have porosity in their bulk material and thus exhibited gas storage properties and thus expected as gaseous fuel loading materials [6, 112]. Various lanthanide and transition metal-lanthanide complexes have been found to have excellent luminescence properties [113]. Yu et al. reported the Zn(II) complex of salen type ligand exhibiting blue photoluminescence with brightness of around 37.2 cd m^{-2} [114]. The LED material also showed excellent thermal stability and thin film coating property. Ni(II), Pd(II) and Pt(II) complexes of salphen derivatives have also shown LED uses [115, 116]. Cu(II) and Zn(II) complexes of thiophene analogue of salphen have been reported as semiconducting material for field-effect transistor with excellent hole mobilities [37]. Thiophene capped salen-metal (V, Ni and Cu) complexes, Pietrangelo et al., where electrochemically polymerised as thin film to get conducting polymers. The polymerised complex materials exhibited enhanced nonlinear optical properties [36].

4.4 Biological activities

Metallosalens exhibits many biological activities as antimicrobial activity, antioxidant activity [117] and anticancer propensity [118]. Their numerous applications have been seen in therapeutics and as biosensors. It has been found that the metal salen have functional enzyme mimic models as superoxide dismutase [119, 120], and Galactose oxidase mimics [121], Cytochrome P-450 mimics [122], Cytochrome

P-450 mimics [123], vitamin B_{12} [124, 125]. Metallosalens are capable of inducing specific damage to DNA or RNA and have been recommended as footprinting agents [126, 127]. Salen complexes are versatile (biomimetic) catalysts for important organic transformations. Derivatives of diaryl-substituted amines linked with metal attached with salen as ligand were experimented in number of cancerous cell lines [128]. Aromatic ring substitution and structural orientation of salen complexes predict the cytotoxicity. Two labile titanium-salen complexes of *cis* configuration were discovered as antitumor agents due to its chelating ability as found in cis-platin [129, 130].

4.5 Sensors

Metal salen complexes have shown the sensory properties for verities of metal ions and small molecules [2, 38]. Colorimetric and fluoremetric both types of responses have been observed depending on the sensor and sensing ions. Chan et al. reported the Pt(II)-salphen based polymeric sensors for the detection of Pd(II), Cd(II), Hg(II), Zn(II), Mg(II), Ca(II), Li(I) and K(I) ions [131, 132]. Wezenberg et al. reported Zn(II)-salphen complexes as metal ion sensors based on demetalation of complexes [133, 134]. Many multimetallic salen complexes have found to be potential sensory properties [2]. Song et al. reported chiral salen based fluorescent polymeric sensor for the enantioselective detection of α-hydroxy carboxylic acids showing fluorescence quenching upon reaction [135]. The same group reported another chiral salen based fluorescent polymeric sensor for the detection of Zn(II) ion as turn-on fluorescence response [136]. Salen based chemosensors for the detection of Al(III) ion based on transmetalation mechanism have also been reported [137].

5. Conclusions

Researcher aims to design or synthesise a molecule with multidirectional use, for developing such a molecule endless work is needed with clarity of innovation leading to novelty. Salen is among those important creation, nevertheless molecule has unimaginable and multiple scope of application ranging from catalysis to biological activities, or as therapeutic use in many medicinal drugs. Salen and its derivatives have been extensively studied because the structural configuration of complex felicitates its importance in various chemical reactions. Widespread use enhances its reliability as catalyst in oxidation, reduction, asymmetric synthesis and many more. The nonsymmetrical salen derivatives have signify to be essential for the preparation of different polymer-supported catalysts that show improved properties (higher activities, catalyst recycling) as collate with parent mono-nuclear complexes. Metallic interference adhere tremendous approach in chemical reaction, presence of metallic centres promotes many specific reaction. Henry reaction, Mannich reaction, Diels-Alder reaction, alkene epoxidation and many such reactions encountered frequently employing salen as transitional part between reactant and product. Metal organic framework (MOF) using salen ligand is recent advancement in the field of macromolecule i.e., supramolecular structure attracting great attention in the field of catalysis and material science. Thus, it is assumed that in near future salen can escort a bloom in the field of catalysis, magnetism, sensors, medicinal areas and material sciences.

DOI: http://dx.doi.org/10.5772/intechopen.88593

Author details

Ashish K. Asatkar[1*], Mamta Tripathi[2] and Deepali Asatkar[3]

1 Department of Chemistry, Government Gundadhur P.G. College, Kondagaon, CG, India

2 School of Studies in Chemistry, Pt. Ravishankar Shukla University, Raipur, CG, India

3 Department of Chemistry, Rani Durgavati University, Jabalpur, MP, India

*Address all correspondence to: ashu.asatkar@gmail.com

References

[1] Pfeiffer P, Breith E, Lubbe E, Tsumaki T. Tricyclische orthokondensierte Nebenvalenzringe. Justus Liebigs Annalen der Chemie. 1933;**503**:84

[2] Clarke RM, Storr T. The chemistry and applications of multimetallic salen complexes. Dalton Transactions. 2014;**43**:9380

[3] Atwood DA, Harvey MJ. Group 13 compounds incorporating salen ligands. Chemical Reviews. 2001;**101**:37

[4] Karmakar M, Chattopadhyay S. A comprehensive overview of the orientation of tetradentate N_2O_2 donor Schiff base ligands in octahedral complexes of trivalent 3d metals. Journal of Molecular Structure. 2019;**1186**:155

[5] Pessoa JC, Correia I. Salan vs. salen metal complexes in catalysis and medicinal applications: Virtues and pitfalls. Coordination Chemistry Reviews. 2019;**388**:227

[6] Yuan G, Jiang H, Zhang L, Liu Y, Cui Y. Metallosalen-based crystalline porous materials: Synthesis and property. Coordination Chemistry Reviews. 2019;**378**:483

[7] Baleizão C, Garcia H. Chiral salen complexes: An overview to recoverable and reusable homogeneous and heterogeneous catalysts. Chemical Reviews. 2006;**106**:3987

[8] Cozzi PG. Metal–salen Schiff base complexes in catalysis: Practical aspects. Chemical Society Reviews. 2004;**33**:410

[9] Bennani YL, Hanessian S. Trans-1,2-diaminocyclohexane derivatives as chiral reagents, scaffolds, and ligands for catalysis: Applications in asymmetric synthesis and molecular recognition. Chemical Reviews. 1997;**97**:3161

[10] Lucet D, Le Gall T, Mioskowski C. The chemistry of vicinal diamines. Angewandte Chemie, International Edition. 1998;**37**:2580

[11] Che C-M, Huang J-S. Metal complexes of chiral binaphthyl Schiff-base ligands and their application in stereoselective organic transformations. Coordination Chemistry Reviews. 2003;**242**:97

[12] Nishikori H, Katsuki T. Catalytic and highly enantioselective aziridination of styrene derivatives. Tetrahedron Letters. 1996;**37**:9245

[13] Belokon Y, Moscalenko M, Ikonnikov N, Yashkina L, Antonov D, Vorontsov E, et al. Asymmetric trimethylsilylcyanation of benzaldehyde catalyzed by (salen) Ti(IV) complexes derived from (R)- and/or (S)-4-hydroxy-5-formyl[2.2] paracyclophane and diamines. Tetrahedron: Asymmetry. 1997;**8**:3245

[14] Cort AD, Mandolini L, Pasquini C, Schiaffino L. Inherently chiral uranyl-salophen macrocycles: Computer-aided design and resolution. The Journal of Organic Chemistry. 2005;**70**:9814

[15] Ciogli A, Cort AD, Gasparrini F, Lunazzi L, Mandolini L, Mazzanti A, et al. Enantiomerization of chiral uranyl–salophen complexes via unprecedented ligand hemilability: Toward configurationally stable derivatives. The Journal of Organic Chemistry. 2008;**73**:6108

[16] Jacobsen EN, Pfaltz A, Yamamoto H. Comprehensive Asymmetric Catalysis. Vol. 2. Berlin, Germany: Springer-Verlag; 1999

[17] Katsuki T. Some recent advances in metallosalen chemistry. Synlett. 2003;**3**:281

[18] Kokubo C, Katsuki T. Highly enantioselective catalytic oxidation

of alkyl aryl sulfides using Mn-salen catalyst. Tetrahedron. 1996;**52**:13895

[19] Fujisaki J, Matsumoto K, Matsumoto K, Katsuk T. Catalytic asymmetric oxidation of cyclic dithioacetals: Highly diastereo- and enantioselective synthesis of the S-oxides by a chiral aluminum(salalen) complex. Journal of the American Chemical Society. 2011;**133**:56

[20] Kleij AW. Nonsymmetrical salen ligands and their complexes: Synthesis and applications. European Journal of Inorganic Chemistry. 2009;**2**:193

[21] Breinbauer R, Jacobsen EN. Cooperative asymmetric catalysis with dendrimeric [Co(salen)] complexes. Angewandte Chemie, International Edition. 2000;**39**:3604

[22] Sellner H, Karjalainen JK, Seebach D. Preparation of dendritic and non-dendritic styryl-substituted salens for cross-linking suspension copolymerization with styrene and multiple use of the corresponding Mn and Cr complexes in enantioselective epoxidations and hetero-Diels–Alder reactions. Chemistry—A European Journal. 2001;7:2873

[23] Kim G-J, Shin J-H. Application of new unsymmetrical chiral Mn(III), Co(II,III) and Ti(IV) salen complexes in enantioselective catalytic reactions. Catalysis Letters. 1999;**63**:83

[24] Renehan MF, Schanz H-J, McGarrigle EM, Dalton CT, Daly AM, Gilheany DG. Unsymmetrical chiral salen Schiff base ligands. Journal of Molecular Catalysis A. 2005;**231**:205

[25] Atkins R, Brewer G, Kokot E, Mockler GM, Sinn E. Copper(II) and nickel(II) complexes of unsymmetrical tetradentate Schiff base ligands. Inorganic Chemistry. 1985;**24**:127

[26] Huber A, Müller L, Elias H, Klement R, Valko M. Cobalt(II) complexes with substituted salen-type ligands and their dioxygen affinity in N,N-dimethylformamide at various temperatures. European Journal of Inorganic Chemistry. 2005;**1459**

[27] Boghaei DM, Mohebi S. Non-symmetrical tetradentate vanadyl Schiff base complexes derived from 1,2-phenylene diamine and 1,3-naphthalene diamine as catalysts for the oxidation of cyclohexene. Tetrahedron. 2002;**58**:5357

[28] Konsler RG, Karl J, Jacobsen EN. Cooperative asymmetric catalysis with dimeric salen complexes. Journal of the American Chemical Society. 1998;**120**:10780

[29] Mazet C, Jacobsen EN. Dinuclear {(salen)Al} complexes display expanded scope in the conjugate cyanation of α,β-unsaturated imides. Angewandte Chemie, International Edition. 2008;**47**:1762

[30] Daly AM, Dalton CT, Renehan MF, Gilheany DG. Enantioselective rhodium catalyzed hydroboration of olefins using chiral bis(aminophosphine) ligands. Tetrahedron Letters. 1999;**40**:3617

[31] Campbell EJ, Nguyen ST. Unsymmetrical salen-type ligands: High yield synthesis of salen-type Schiff bases containing two different benzaldehyde moieties. Tetrahedron Letters. 2001;**42**:1221

[32] Annis DA, Jacobsen EN. Polymer-supported chiral co(salen) complexes: Synthetic applications and mechanistic investigations in the hydrolytic kinetic resolution of terminal epoxides. Journal of the American Chemical Society. 1999;**121**:4147

[33] Anyanwu UK, Venkataraman D. Effect of spacers on the activity of soluble polymer supported catalysts for the asymmetric addition of diethylzinc

to aldehydes. Tetrahedron Letters. 2003;**44**:6445

[34] Rigamonti L, Demartin F, Forni A, Righetto S, Pasini A. Copper(II) complexes of salen analogues with two differently substituted (push-pull) salicylaldehyde moieties. A study on the modulation of electronic asymmetry and nonlinear optical properties. Inorganic Chemistry. 2006;**45**:10976

[35] Dabelstein W, Reglitzky A, Schutze A, Reders K. Automotive fuels. In: Ullmann's Encyclopedia of Industrial Chemistry. Weinheim: Wiley-VCH; 2007

[36] Pietrangelo A, Sih BC, Boden BN, Wang Z, Li Q, Chou KC, et al. Nonlinear optical properties of Schiff-base-containing conductive polymer films electro-deposited in microgravity. Advanced Materials. 2008;**20**:2280

[37] Asatkar AK, Senanayak SP, Bedi A, Panda S, Narayan KS, Zade SS. Zn(II) and Cu(II) complexes of a new thiophenebased salphen-type ligand: Solution-processable high-performance field-effect transistor materials. Chemical Communications. 2014;**50**:7036

[38] Whiteoak CJ, Salassa G, Kleij AW. Recent advances with p-conjugated salen systems. Chemical Society Reviews. 2012;**41**:622

[39] Kuo K-L, Huang C-C, Lin Y-C. Synthesis and photophysical properties of multinuclear zinc-salophen complexes: Enhancement of fluorescence by fluorene termini. Dalton Transactions. 2008;**3889**

[40] Kleij AW, Tooke DM, Kuil M, Lutz M, Spek AL, Reek JNH. ZnII–salphen complexes as versatile building blocks for the construction of supramolecular box assemblies. Chemistry—A European Journal. 2005;**11**:4743

[41] Ada'n ECE, Belmonte MM, Martin E, Salassa G, Buchholz JB, Kleij AW. A short desymmetrization protocol for the coordination environment in bis-salphen scaffolds. The Journal of Organic Chemistry. 2011;**76**:5404

[42] Kleij AW. New templating strategies with salen scaffolds (salen=N,N'-bis(salicylidene)ethylenediamine dianion). Chemistry—A European Journal. 2008;**14**:10520

[43] Castilla AM, Curreli S, Belmonte MM, Ada'n ECE, Buchholz JB, Kleij AW. Modular synthesis of heterobimetallic salen structures using metal templation. Organic Letters. 2009;**11**:5218

[44] Akine S, Taniguchi T, Nabeshima T. Synthesis and crystal structure of a novel triangular macrocyclic molecule, tris(H_2saloph), and its water complex. Tetrahedron Letters. 2001;**42**:8861

[45] Akine S, Nabeshima T. Cyclic and acyclic oligo(N_2O_2) ligands for cooperative multi-metal complexation. Dalton Transactions. 2009:10395

[46] Jiang J, MacLachlan MJ. Unsymmetrical triangular Schiff base macrocycles with cone conformations. Organic Letters. 2010;**12**:1020

[47] Frischmann PD, Jiang J, Hui JK-H, Grzybowski JJ, MacLachlan MJ. Reversible-irreversible approach to Schiff base macrocycles: Access to isomeric macrocycles with multiple salphen pockets. Organic Letters. 2008;**10**:1255

[48] Furukawa H, Cordova KE, O'Keeffe M, Yaghi OM. The chemistry and applications of metal-organic frameworks. Science. 2013;**341**:1230444

[49] Liu TF, Feng D, Chen YP, Zou L, Bosch M, Yuan S, et al. Topology-guided design and syntheses of highly stable

mesoporous porphyrinic zirconium metal–organic frameworks with high surface area. Journal of the American Chemical Society. 2015;**137**:413

[50] Mo K, Yang Y, Cui Y. A homochiral metal–organic framework as an effective asymmetric catalyst for cyanohydrin synthesis. Journal of the American Chemical Society. 2014;**136**:1746

[51] Shultz AM, Farha OK, Adhikari D, Sarjeant AA, Hupp JT, Nguyen ST. Selective surface and near-surface modification of a noncatenated, catalytically active metal-organic framework material based on Mn(salen) struts. Inorganic Chemistry. 2011;**50**:3174

[52] Shultz AM, Sarjeant AA, Farha OK, Hupp JT, Nguyen ST. Post-synthesis modification of a metal–organic framework to form metallosalen-containing MOF materials. Journal of the American Chemical Society. 2011;**133**:13252

[53] Song F, Wang C, Falkowski JM, Ma L, Lin W. Isoreticular chiral metal-organic frameworks for asymmetric alkene epoxidation: Tuning catalytic activity by controlling framework catenation and varying open channel sizes. Journal of the American Chemical Society. 2010;**132**:15390

[54] Jeon Y-M, Armatas GS, Heo J, Kanatzidis MG, Mirkin CA. Amorphous infinite coordination polymer microparticles: A new class of selective hydrogen storage materials. Advanced Materials. 2008;**20**:2105

[55] Roesky PW, Bhunia A, Lan Y, Powell AK, Kureti S. Salen-based metal–organic frameworks of nickel and the lanthanides. Chemical Communications. 2011;**47**:2035

[56] Cort AD, Murua JIM, Pasquini C, Pons M, Schiaffino L. Evaluation of chiral recognition ability of a novel uranyl–salophen-based receptor: An easy and rapid testing protocol. Chemistry—A European Journal. 2004;**10**:3301

[57] Bera MK, Chakraborty C, Malik S. How the stereochemistry decides the selectivity: An approach towards metal ion detection. New Journal of Chemistry. 2015;**39**:9207

[58] Salassa G, Coenen MJJ, Wezenberg SJ, Hendriksen BLM, Speller S, Elemans JAAW, et al. Extremely strong self-assembly of a bimetallic salen complex visualized at the single-molecule level. Journal of the American Chemical Society. 2012;**134**:7186

[59] Leoni L, Cort AD. The supramolecular attitude of metal–salophen and metal–salen complexes. Inorganics. 2018;**6**(42):1

[60] Akine S, Taniguchi T, Nabeshima T. Helical metallohost–guest complexes via site-selective transmetalation of homotrinuclear complexes. Journal of the American Chemical Society. 2006;**128**:15765

[61] Dutta PK, Asatkar AK, Zade SS, Panda S. Oxidative addition of disulfide/diselenide to group 10 metal(0) and in situ functionalization to form neutral thiasalen/selenasalen group 10 metal(II) complexes. Dalton Transactions. 2014;**43**:1736

[62] Panda S, Krishna GR, Reddy CM, Zade SS. Synthesis, characterization and coordination properties of bis(alkyl)selenosalen ligands. Dalton Transactions. 2011;**40**:6684

[63] Dutta PK, Panda S, Krishna GR, Reddy CM, Zade SS. Reaction time dependent formation of Pd(II) and Pt(II) complexes of bis(methyl) thiasalen podand. Dalton Transactions. 2013;**42**:476

[64] Nguyen QT, Jeong JH. Synthesis and X-ray structure of a Cu(II) complex

of N,N'-bis(2-pyridylmethylidene)-(R,R)-1,2-diaminocyclohexane and its catalytic application for asymmetric Henry reaction. Polyhedron. 2006;**25**:1787-1790

[65] Asatkar AK, Panda S, Zade SS. Thiophene-based salen-type new ligands, their structural aspects and a dimeric Cu(II) complex. Polyhedron. 2015;**96**:25

[66] Berube CD, Gambarotta S, Yap GPA, Cozzi PG. Di- and trivalent dinuclear samarium complexes supported by pyrrole-based tetradentate Schiff bases. Organometallics. 2003;**22**:434

[67] Ooi T, Miura T, Maruoka K. Highly efficient, catalytic Meerwein–Ponndorf–Verley reduction with a novel bidentate aluminum catalyst. Angewandte Chemie, International Edition in English. 1998;**37**:2347

[68] Ooi T, Itagaki Y, Miura T, Maruoka K. Simultaneous functional group manipulation in the Meerwein-Ponndorf-Verley reduction process catalyzed by bidentate aluminum reagent. Tetrahedron Letters. 1999;**40**:2137

[69] Osamura Y, Terada K, Kobayashi Y, Okazaki R, Ishiyama Y. A molecular orbital study of the mechanism of chlorination reaction of benzene catalyzed by Lewis acid. Journal of Molecular Structure. 1999;**461-462**:399

[70] Berberich H, Roesky PW. Homoleptic lanthanide amides as homogeneous catalysts for the Tishchenko reaction. Angewandte Chemie, International Edition in English. 1998;**37**:1569

[71] Santelli M, Pons J-M. Lewis Acids and Selectivity in Organic Synthesis. New York: CRC Press; 1996

[72] Saito S, Shiozawa M, Ito M, Yamamoto H. Conceptually new directed Aldol condensation using aluminum tris(2,6-diphenylphenoxide). Journal of the American Chemical Society. 1998;**120**:813

[73] Nelson SG, Peelen TJ, Wan Z. Mechanistic alternatives in Lewis acid-catalyzed acyl halide aldehyde cyclocondensations. Tetrahedron Letters. 1999;**40**:6541

[74] Yoon TP, Dong VM, MacMillan DWC. Development of a new Lewis acid-catalyzed Claisen rearrangement. Journal of the American Chemical Society. 1999;**121**:9726

[75] Foley SR, Yap GPA, Richeson DS. Formation of novel tetrasulfido tin complexes and their ability to catalyze the cyclotrimerization of aryl isocyanates. Organometallics. 1999;**18**:4700

[76] Dalton CT, Ryan KM, Wall VM, Bousquet C, Gilheany DG. Recent progress towards the understanding of metal–salen catalysed asymmetric alkene epoxidation. Topics in Catalysis. 1998;**5**:75

[77] Achard TRJ, Clutterbuck LA, North M. Asymmetric catalysis of carbon-carbon bond-forming reactions using metal(salen) complexes. Synlett. 2005;**12**:1828

[78] Venkataramanan NS, Kuppuraj G, Rajagopal S. Metal–salen complexes as efficient catalysts for the oxygenation of heteroatom containing organic compounds—Synthetic and mechanistic aspects. Coordination Chemistry Reviews. 2005;**249**:1249

[79] Higuchi Y, Yagi T, Yasuoka N. Unusual ligand structure in Ni-Fe active center and an additional Mg site in hydrogenase revealed by high resolution X-ray structure analysis. Structure. 1997;**5**:1671

[80] Higuchi Y, Ogata H, Miki K, Yasuoka N, Yagi T. Removal of the bridging ligand atom at the Ni-Fe

active site of [NiFe] hydrogenase upon reduction with H_2, as revealed by X-ray structure analysis at 1.4 Å resolution. Structure. 1999;7:549

[81] Spencer DJE, Marr AC, Schrçder M. Structural mimics for the active site of [NiFe] hydrogenase. Coordination Chemistry Reviews. 2001;**219-221**:1055

[82] Volbeda A, Charon MH, Piras C, Hatchikian EC, Frey M, Camps JCF. Crystal structure of the nickel–iron hydrogenase from *Desulfovibrio gigas*. Nature. 1995;**373**:580

[83] Svetlitchnyi V, Dobbek H, Klaucke WM, Meins T, Thiele B, Romer P, et al. A functional Ni-Ni-[4Fe-4S] cluster in the monomeric acetyl-CoA synthase from *Carboxydothermus hydrogenoformans*. Proceedings of the National Academy of Sciences of the United States of America. 2004;**101**:446

[84] Darnault C, Volbeda A, Kim EJ, Legrand P, Vernede X, Lindahl PA, et al. Ni-Zn-[Fe4-S4] and Ni-Ni-[Fe4-S4] clusters in closed and open subunits of acetyl-CoA synthase/carbon monoxide dehydrogenase. Nature Structural Biology. 2003;**10**:271

[85] Doukov TI, Iverson TM, Seravalli J, Ragsdale SW, Drennan CL. A Ni-Fe-Cu center in a bifunctional carbon monoxide dehydrogenase/acetyl-CoA synthase. Science. 2002;**298**:567

[86] Drennan CL, Heo JY, Sintchak MD, Schreiter E, Ludden PW. Life on carbon monoxide: X-ray structure of *Rhodospirillum rubrum* Ni-Fe-S carbon monoxide dehydrogenase. Proceedings of the National Academy of Sciences of the United States of America. 2001;**98**:11973

[87] Dobbek H, Svetlitchnyi V, Gremer L, Huber R, Meyer O. Crystal structure of a carbon monoxide dehydrogenase reveals a [Ni-4Fe-5S] cluster. Science. 2001;**293**

[88] Ermler U, Grabarse W, Shima S, Goubeaud M, Thauer RK. Crystal structure of methyl-coenzyme M reductase: The key enzyme of biological methane formation. Science. 1997;**278**:1457

[89] Mirkhani V, Tangestaninejad S, Moghadam M, BaltorkIPM, KargarH. Efficientoxidation of sulfides with sodium periodate catalyzed by manganese(III) Schiff base complexes. Journal of Molecular Catalysis A: Chemical. 2005;**242**:251

[90] Shibasaki M, Kanai M, MatsunagaS, KumagaiN. Recentprogress in asymmetric bifunctional catalysis using multimetallic systems. Accounts of Chemical Research. 2009;**42**:1117

[91] Walsh PJ, Kozlowski MC. Fundamentals of Asymmetric Catalysis. Sausalito: University Science Books; 2009

[92] Pozzi G, Shepperson I. Fluorous chiral ligands for novel catalytic systems. Coordination Chemistry Reviews. 2003;**242**:115

[93] Canali L, Sherrington DC. Utilisation of homogeneous and supported chiral metal(salen) complexes in asymmetric catalysis. Chemical Society Reviews. 1999;**28**:85

[94] Zulauf A, Mellah M, Hong X, Schulz E. Recoverable chiral salen complexes for asymmetric catalysis: Recent progress. Dalton Transactions. 2010;**39**:6911

[95] Leung ACW, MacLachlan M. Schiff Base complexes in macromolecules. Journal of Inorganic and Organometallic Polymers and Materials. 2007;**17**:57

[96] Madhavan N, Jones CW, Weck M. Rational approach

to polymer-supported catalysts: Synergy between catalytic reaction mechanism and polymer design. Accounts of Chemical Research. 2008;**41**:1153

[97] Handa S, Gnanadesikan V, Matsunaga S, Shibasaki M. Heterobimetallic transition metal/rare earth metal bifunctional catalysis: A Cu/Sm/Schiff base complex for Syn-selective catalytic asymmetric nitro-Mannich reaction. Journal of the American Chemical Society. 2010;**132**:4925

[98] Handa S, Nagawa K, Sohtome Y, Matsunaga S, Shibasaki M. A heterobimetallic Pd/La/Schiff base complex for anti-selective catalytic asymmetric nitroaldol reactions and applications to short syntheses of β-adrenoceptor agonists. Angewandte Chemie, International Edition. 2008;**47**:3230

[99] Miller JS, Calabrese JC, Epstein AJ, Bigelow RW, Zhang JH, Reiff WM. Ferromagnetic properties of one-dimensional decamethylferrocenium tetracyanoethylenide (1:1):[Fe(η5-C5Me5)2]$^{\bullet+}$[TCNE]$^{\bullet}$. Journal of the Chemical Society, Chemical Communications. 1986;**1026**

[100] Miller JS, Calabrese JC, Rommelmann H, Chittipeddi SR, Zhang JH, Reiff WM, et al. Ferromagnetic behavior of [Fe(C5Me5)2]$^{\bullet+}$ [TCNE]$^{\bullet-}$. Structural and magnetic characterization of decamethylferrocenium tetracyanoethenide, [Fe(C5Me5)2]$^{\bullet+}$ [TCNE]$^{\bullet-}$•MeCN and decamethylferrocenium pentacyanopropenide, [Fe(C5Me5)2]$^{\bullet+}$ [C3(CN)5]$^{-}$. Journal of the American Chemical Society. 1987;**109**:769

[101] Glaser T. Exchange coupling mediated by extended phloroglucinol ligands: Spin-polarization vs. heteroradialene-formation. Coordination Chemistry Reviews. 2013;**257**:140

[102] Glaser T, Gerenkamp M, Fröhlich R. Targeted synthesis of erromagnetically coupled complexes with modified 1,3,5-trihydroxybenzene ligands. Angewandte Chemie, International Edition. 2002;**41**:3823

[103] Glaser T, Heidemeier M, Grimme S, Bill E. Targeted ferromagnetic coupling in a trinuclear copper(II) complex: Analysis of the St = 3/2 spin ground state. Inorganic Chemistry. 2004;**43**:5192

[104] Glaser T, Heidemeier M, Fröhlich R, Hildebrandt P, Bothe E, Bill E. Trinuclear nickel complexes with triplesalen ligands: Simultaneous occurrence of mixed valence and valence tautomerism in the oxidized species. Inorganic Chemistry. 2005;**44**:5467

[105] Ratera I, Veciana J. Playing with organic radicals as building blocks for functional molecular materials. Chemical Society Reviews. 2012;**41**:303

[106] Theil H, von Richthofen C-GF, Stammler A, Bögge H, Glaser T. Ferromagnetic coupling by the spin-polarization mechanism in a trinuclear VIV triplesalen complex. Inorganica Chimica Acta. 2008;**361**:916

[107] Mukherjee C, Stammler A, Bögge H, Glaser T. Trinuclear C3-symmetric extension of Jacobsen's catalyst: Synthesis, characterization, and catalytic properties of a chiral trinuclear MnIII triplesalen complex. Inorganic Chemistry. 2009;**48**:9476

[108] Mukherjee C, Stammler A, Bögge H, Glaser T. Do trinuclear triplesalen complexes exhibit cooperative effects? Synthesis, characterization, and enantioselective catalytic sulfoxidation by chiral trinuclear FeIII triplesalen complexes. Chemistry—A European Journal. 2010;**16**:10137

[109] Theil H, von Richthofen C-G, Stammler A, Stammler A,

Bögge H, Glaser T. From triplesalen to triplesalophen: Ferromagnetic interactions through spin-polarization in a trinuclear Ni-II triplesalophen complex. European Journal of Inorganic Chemistry. 2011;**49**

[110] Glaser T, Heidemeier M, Strautmann JBH, Bögge H, Stammler A, Krickemeyer E, et al. Trinuclear copper complexes with triplesalen ligands: Geometric and electronic effects on ferromagnetic coupling via the spin-polarization mechanism. Chemistry—A European Journal. 2007;**13**:9191

[111] Wezenberg SJ, Kleij AW. Material applications for Salen frameworks. Angewandte Chemie, International Edition. 2008;**47**:2354

[112] Crane AK, MacLachlan MJ. Portraits of porosity: Porous structures based on metal salen complexes. European Journal of Inorganic Chemistry. 2012;**17**

[113] Yang X, Jones RA, Huang S. Luminescent 4f and d-4f polynuclear complexes and coordination polymers with flexible salen-type ligands. Coordination Chemistry Reviews. 2014;**63**:273-274

[114] Yu G, Liu Y, Song Y, Wu X, Zhu D. A new blue light-emitting material. Synthetic Metals. 2001;**117**:211

[115] Lavastre O, Illitchev I, Jegou G, Dixneuf PH. Discovery of new fluorescent materials from fast synthesis and screening of conjugated polymers. Journal of the American Chemical Society. 2002;**124**:5278

[116] Che C-M, Kwok C-C, Lai S-W, Rausch AF, Finkenzeller WJ, Zhu N, et al. Photophysical properties and OLED applications of phosphorescent platinum(II) Schiff Base complexes. Chemistry—A European Journal. 2010;**16**:233

[117] Iranzo O. Manganese complexes displaying superoxide dismutase activity: A balance between different factors. Bioorganic Chemistry. 2011;**39**:73

[118] Erxleben A. Transition metal salen complexes in bioinorganic and medicinal chemistry. Inorganica Chimica Acta. 2018;**472**:40

[119] Barondeau DP, Kassmann CJ, Bruns CK, Tainer JA, Getzoff ED. Nickel superoxide dismutase structure and mechanism. Biochemistry. 2004;**43**:8038

[120] Wuerges J, Lee J-W, Yim YI, Yim H-S, Kang S-O, Djinovic Carugo K. Crystal structure of nickel-containing superoxide dismutase reveals another type of active site. Proceedings of the National Academy of Sciences of the United States of America. 2004;**101**:8569

[121] Lyons CTL, Stack TDP. Recent advances in phenoxyl radical complexes of salen-type ligands as mixed-valent galactose oxidase models. Coordination Chemistry Reviews. 2013;**257**:528

[122] Cho K-B, Nam W, Hirao H, Shaik S. To rebound or dissociate? This is the mechanistic question in C–H hydroxylation by heme and nonheme metal–oxo complexes. Chemical Society Reviews. 2016;**45**:1197

[123] Wollenweber E, Harborne JB, Mabry TJ, editors. Flavonoids: Advances in Research. London, New York: Chapman & Hall; 1982

[124] Tsou TT, Loots M, Halpern J. Kinetic determination of transition metal-alkyl bond dissociation energies: Application to organocobalt compounds related to B12 coenzymes. Journal of the American Chemical Society. 1982;**104**:623

[125] Halpern J, Ng FTT, Rempel GL. Metal-alkyl bond dissociation energies

in organocobalt compounds related to vitamin B12 coenzymes. Journal of the American Chemical Society. 1979;**101**:7124

[126] Herchel R, Sindelar Z, Travnicek Z, Zboril R, Vanco J. Novel 1D chain Fe(III)-salen-like complexes involving anionic heterocyclic N-donor ligands. Synthesis, X-ray structure, magnetic, ^{57}Fe Mössbauer, and biological activity studies. Dalton Transactions. 2009:9870

[127] Asatkar AK, Tripathi M, Panda S, Pande R, Zade SS. Cu(I) complexes of bis(methyl)(thia/selena) salen ligands: Synthesis, characterization, redox behavior and DNA binding studies. Spectrochimica Acta, Part A: Molecular and Biomolecular Spectroscopy. 2017;**171**:18

[128] Saini AK, Kumari P, Sharma V, Mathur P, Mobin SM. Varying structural motifs in the salen based metal complexes of Co(II), Ni(II) and Cu(II): Synthesis, crystal structures, molecular dynamics and biological activities. Dalton Transactions. 2016;**45**:19096

[129] Gust R, Ott I, Posselt D, Sommer K. Development of cobalt(3,4-diarylsalen) complexes as tumor therapeutics. Journal of Medicinal Chemistry. 2004;**47**:5837

[130] Tzubery A, Tshuva EY. Trans titanium(IV) complexes of salen ligands exhibit high antitumor activity. Inorganic Chemistry. 2011;**50**:7946

[131] Guo Z, Tong W-L, Chan MCW. Axially rotating (Pt-salphen)2 phosphorescent coordination frameworks. Chemical Communications. 2009:6189

[132] Sun S, Tong W-L, Chan MCW. Alternating poly(Pt-salphen)-(p-phenyleneethynylene) as phosphorescent conjugated linear-rod and coilable sensory materials. Macromolecular Rapid Communications. 2010;**31**:1965

[133] Wezenberg SJ, Escudero-Ada'n EC, Anselmo D, Buchholz JB, Kleij AW. Dimetallic activation of dihydrogen phosphate by Zn (salphen) chromophores. European Journal of Inorganic Chemistry. 2010;**4611**

[134] Wezenberg SJ, Ada'n ECE, Buchholz JB, Kleij AW. Colorimetric discrimination between important alkaloid nuclei mediated by a bis-salphen chromophore. Organic Letters. 2008;**10**:3311

[135] Song F, Wei G, Wang L, Jiao J, Cheng Y, Zhu C. Salen-based chiral fluorescence polymer sensor for enantioselective recognition of α-hydroxyl carboxylic acids. The Journal of Organic Chemistry. 2012;**77**(10):4759

[136] Song F, Ma X, Hou J, Huang X, Cheng Y, Zhu C. (R,R)-salen/salan-based polymer fluorescence sensors for Zn^{2+} detection. Polymer. 2011;**52**:6029

[137] Cheng J, Ma X, Zhang Y, Liu J, Zhou X, Xiang H. Optical chemosensors based on transmetalation of salen-based Schiff base complexes. Inorganic Chemistry. 2014;**53**(6):3210

Chapter 8

Ligands and Coordination Compounds Used as New Photosensitized Materials for the Construction of Solar Cells

Yenny Patricia Avila Torres

Abstract

Ligand with conjugated π systems presents high planar and delocalized electronic density, which allows it to capture the radiations with an energy interval of wavelengths between 400 and 600 nm. The ligands can be linked to inorganic materials favoring the interchange in the system. Lewis acids improve the electronic distribution between donor and acceptor favoring the optical and electronic properties, yielding superior efficiencies. In this chapter, the evolution of the ligands such as Porphyrins, Metal-free organic dyes, and Ruthenium complexes used, and the construction of solar cells is described. In this context, three different small-molecule acceptors-donors are reported; *o*-PDT, *m*-PDT and *p*-PDT, based on phenyldiammine (PD) as spacer, and Thiazole (T) were designed and synthesized. There were estimated electronic, optical and photovoltaic parameters for these molecules. The interaction energies of functional groups for PD and T molecules, with DFT/B3LYP method, gas phase with 6-31g (d, 2p) basis sets, were represented and computed. The best photovoltaic parameters were described for *p*-PDT with PCE 26.18%, Jsc = 14.79 $mAcm^2$ and ΔE = 2.66 eV. The metal ion influences the electronic properties and decreases the ΔE GAP. The incorporation of the transition metals into hyperconjugated systems provides rigidity and effects of electronic back donation.

Keywords: photovoltaic cells, semiconductor, coordination compounds, dye-sensitized

1. Introduction

The current environmental problems and the energy crisis have led to creating new technologies. The renewable energies such as: biofuels, biomass, wind, geothermal, hydraulic, solar, tidal, among others become the main source of the energy generation. The use of solar cells represents an alternative among renewable energies. The development of new materials goes from inorganic structures until polymers, small molecules as organic photovoltaic (OPV) and photosensitized organic materials [1].

In this context, the discovery of ultrafast charge transfer between the semiconductor polymer and inorganic semiconductors have allowed that the OPV have

IntechOpen

passed in a decade of values close to 1% of efficiency until exceeding 10% [2]. This rapid evolution is motivated by its high potential for generating flexible, lightweight and low cost panels changing the classic concept of panels photovoltaic. In the case of the traditional polymers, the electrons are highly localized and require great energy to be excited (>5 eV) converting them into electrical insulators. In contrast, in conjugated polymer and structures the electrons from Z orbitals form π-type bonds that are associated with lower energies, corresponding to the range of ultraviolet and visible radiation. In molecular solids the transition $\pi \rightarrow \pi^*$ that occurs between the occupied molecular orbital of higher energy (HOMO) and the lower energy of orbital (LUMO) determine the equivalent to the forbidden band energy of inorganic semiconductors [3, 4]. On the other hand, they had developed sensitized cell from organic dyes. These are also called Grätzel cells. Photoelectrons that introduce into the conduction band of TiO_2 that works as a semiconductor, under light illumination [5].

Dye-sensitized cell (DSSC) have been developed as functional biomimetic models of biological process. In the nature exists dyes with electronic properties that allows to purpose design news in solar panel. Chlorophyll, constituted an example in where there is light absorption and charge-carrier transport. The organic molecule is coupled to semiconductor enhancing the Gap band. This electronic transfer promote absorption to the visible region, which increase its applications [6].

The researchers in a world context have designed, developed and synthetized ruthenium complexes, porphyrins, metal-free organic dyes and organic molecules in this field.

1.1 Ruthenium complexes

The efficiency of DSSC depends on different requirements listed below [7, 8]:

i. Broad and strong absorption, preferably extending from energies greater than 900 nm.

ii. The dye needs to be photochemically, thermally, and electrochemically robust within the DSSC in order to withstand the harsh conditions of a practical module.

iii. Firm, irreversible adsorption to the semiconductor's surface (TiO_2) and strong electronic coupling between its excited state and the semiconductor conduction band.

iv. Reduction potential is sufficiently higher than the semiconductor conduction band Edge in order to enable charge injection.

v. Chemical stability in the ground and the excited states for rapid dye regeneration and charge-injection processes.

Different types of dyes have been tested in the DSSC setting, including: transition-metal complexes, organic dyes, porphyrins and phthalocyanines [9–12]; however, in terms of photovoltaic performance and long-term stability, Ru(II) complexes comprise the most successful family of DSSCs sensitizers, shown in **Table 1**. A study on these champion dyes reveals that majority are derivate of **N3**. The **N3** dye represents the first high-performance Ru(II) sensitizer reported in 1993 by Grätzel and co-workers [13], affording power conversion efficiencies of 10.3%. The chemical modification of **N3** and **N719** is made possible because only two anchoring groups are necessary for electron injection [14, 15] thus various functional groups can be installed to block the electrolyte from interacting with the surface or absorb more light. Then, the performance of these successful sensitizers encompasses ligands that combine extended π-conjugated systems, aspiring to enhance the optical absorptivity of the semiconductor's surface, along with long hydrophobic alkyl chains, aiming an increase of tolerance against water attack (**Table 1**). Equally importantly **N719**, which essentially differ only in the protonation state of tetra-protonated parent dye **N3**, afford a nearly quantitative conversion of incident photons into electric current over a large spectral range. The improved efficiency of **N719**, was mainly attributed to the increased cell voltage. Since 1993, chemical modifications of these early Ru(II) complexes have led to researchers achieving power conversion efficiencies up to 11.7% (**C106** dye) [16–24], where one of the DCBPy ligands has been replaced with an extended conjugation using thiophenes and long alkyl chains, lastly, these prevents interfacial recombination [25].

1.2 Porphyrins

The so-called solar cells sensitized by a dye are a type of hybrid devices that have reached a higher degree of development so far. Within his field, porphyrins represent a very interesting alternative because there is efficient model harnessing sunlight. These systems can be synthetized in bulk heterojunction (BHJ) organic solar cells. The interaction of macrocycles with metal ions such as: Fe^{2+}, Fe^{3+}, Co^{2+}, Co^{3+}, Ni^{2+}, Zn^{2+}, Cu^{2+}, Ru^{2+}, Pd^{2+} and Pt^{2+} and hydrogen, alkyl, cycloalkyl, cyclohexyl, cycloheptyl, cyclooctyl, haloalkyl, perhaloalkyl, ether chains have permitted the stabilizations of promising new collections. In dye-sensitized solar porphyrin-based push-pull.

photosensitizers have demonstrated their potential as large and rigid planar conjugated structures, which enhance p-electron delocalization and promote intermolecular π-π interaction, as well charge transport in devices. A problem that they can presents is the effect by lack of light-harvesting beyond 850 nm, thus limiting their cell performance. In the papers, it has been reported that 50% of the total solar phonon flux is located in the red and near-infrared spectra. Zhu and colleagues had reported in 2016 that is quite urgent to develop efficient NIR absorbing molecules for high performance organic solar cells. In the next table, the authors show different publications about the development of new bioinspired porphyrin materials (**Table 2**).

1.3 Metal-free organic dyes

The DSSC free organic dyes are sensitized molecules whose perspective are aimed at staking, on top of one other in order to obtain panchromatic absorption. **Table 3** shows azo, cyano, thiophene, and carbonyl with highly conjugated. A PCE value at 14.7 has been reported by Kakiag et al. [49]. The PCE increase with V_{oc} and J_{sc} and the best properties were associated with carboxylic group and highly polarizability in the presence of nitrile group.

The extension of the conjugated chain and the substitution of the thiophene groups do not represent a marked difference that allows concluding a relationship between the photovoltaic properties and the structure.

Therefore, this article reports the bibliographic revision for these compounds, specifying the following parameters: Chemical name, abbreviation, structure, power conversion efficiencies (PCE), J_{sc} (short-circuit-current), V_{oc} (open circuit voltage) electrolyte used and authors.

Chemical name	Author's designation	Structure	PCE (%)	J_{sc} (mA/cm^2)	V_{oc} (mV)	Electrolyte
cis-Bis (isothiocyanato)bis (2,2′-bipyridyl-4,4′-dicarboxylato) ruthenium(II)	**N3**	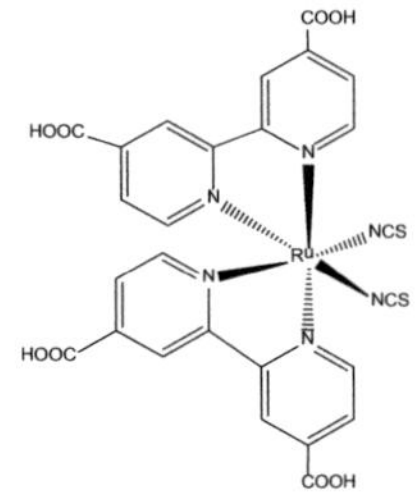	10.3	7.9	660	I_3/I^-
Di-tetrabutylammonium *cis*-bis (isothiocyanato)bis (2,2′-bipyridyl-4,4′-dicarboxylato) ruthenium(II)	**N719**	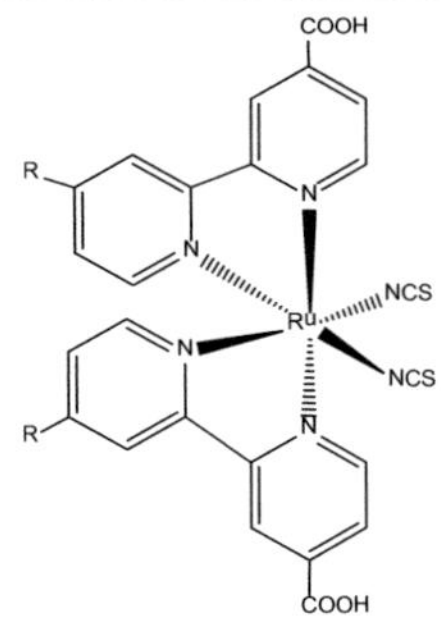	11.18	17.73	846	I_3/I^-
$[(C_4H_9)_4N]_3$ [Ru-(Htcterpy)$(NCS)_3$] (tcterpy) 4,4′,4″-tricarboxy-2,2′,2″-terpyridine	**Black dye**	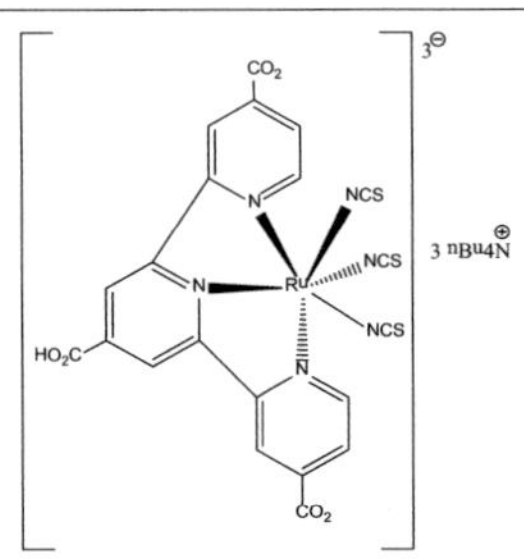	10.4	20	720	I_3/I^-
cis-Ru (4,4′-dicarboxylic acid-2, 2′-bipyridine) (4,4′-dinonyl-2,2′-bipyridine)(NCS)2	**Z907**	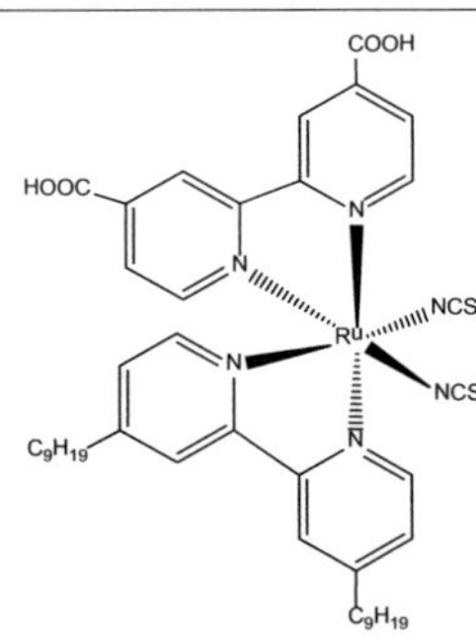	9.5	12.5	730	
cis-Bis(isothiocyanato) (2,2′-bipyridyl-4,4′-dicarboxylato)(4,4′-bis(5-hexylthiophen-2-yl)-2,2′-bipyridyl) ruthenium(II), Ruthenate(2−), [[2,2′-bipyridine]-4,4′-dicarboxylato (2−)-N1,N1′][4,4′-bis(5-hexyl-2-thienyl)-2,2′-bipyridine-N1, N1′]bis(thiocyanato-N)-, hydrogen (1:2)	**C101**	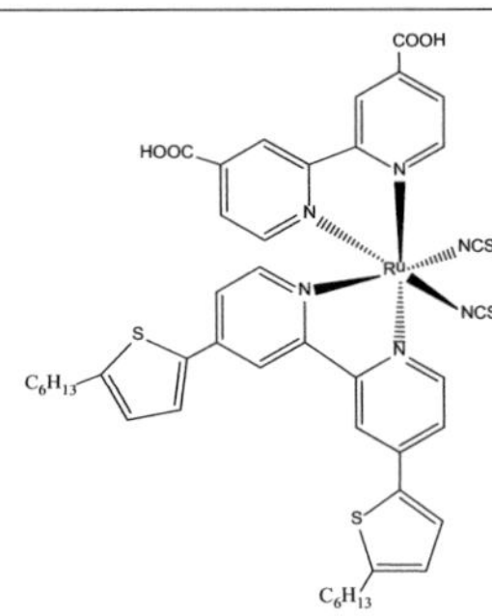	11	17.9	778	I_3/I^-

Chemical name	Author's designation	Structure	PCE (%)	J_{sc} (mA/cm^2)	V_{oc} (mV)	Electrolyte
TBA(Ru[(4-carboxylic acid-4′-carboxylate-2,2′-bipyridine)(Ligand-11)(NCS)$_2$])	**CYC-B11**	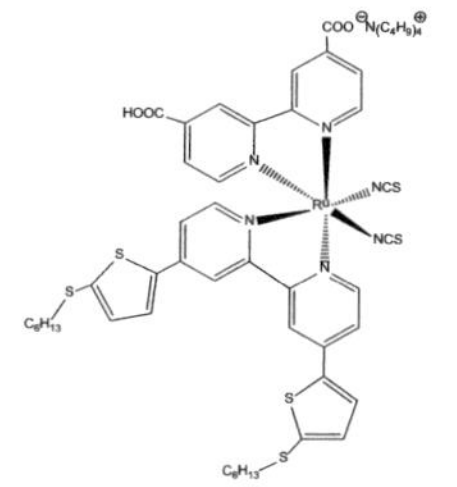	11.5	20.05	743	I_3/I^-
—	**C106**	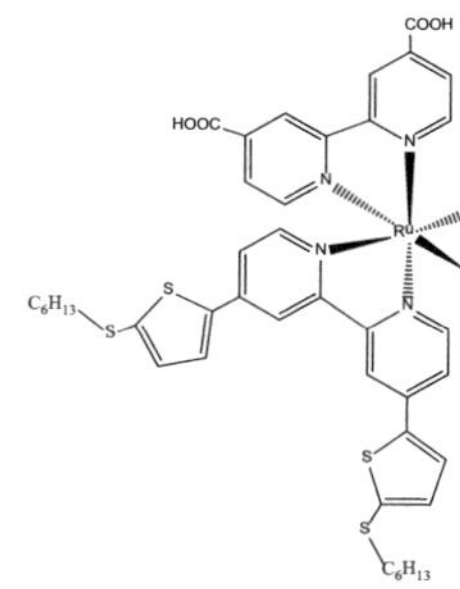	11.7%	19.8	758	I_3/I^-
—	**GS3**	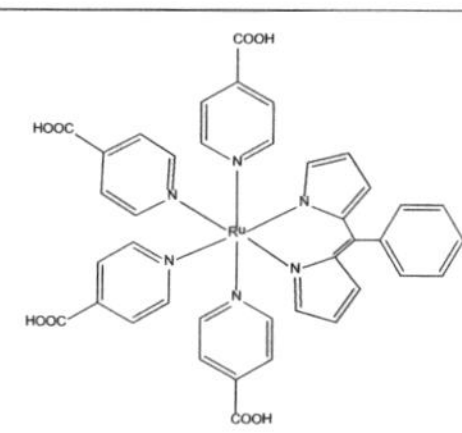	2.79	9.78	435	I_3/I^-
—	**NCSU-10**	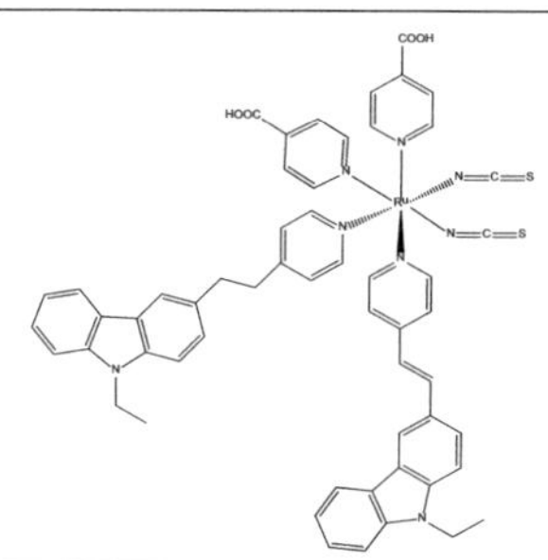	8.34	18.2	703	I_3/I^-
—	**Complex 16**	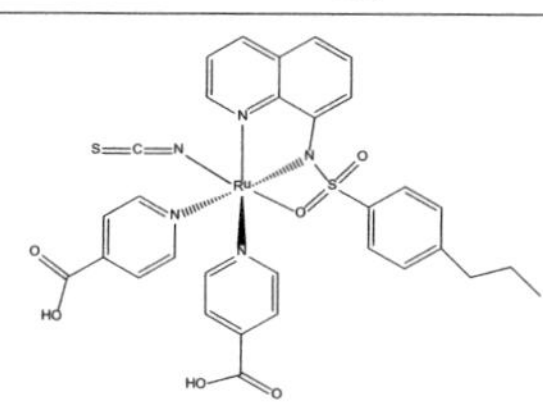	1.26	4.53	496	I_3/I^-
—	**[Ru]2**	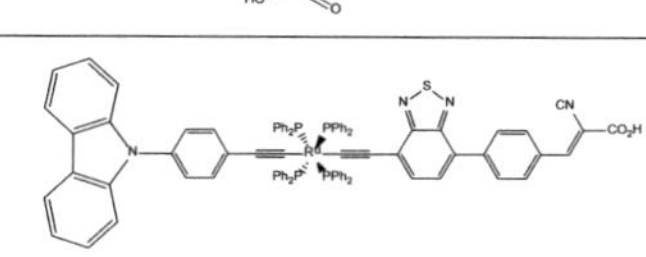	6.45	13.48	650	—
—	**[Ru]3**	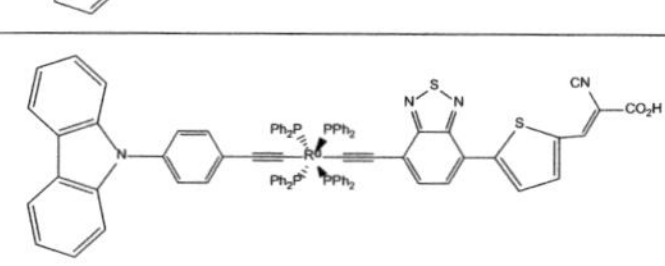	5.23	11.77	598	—

—, it is not mentioned in the article.

Table 1.
Ruthenium complexes for DSSC [26–33].

Chemical name	Author's designation	Structure	PCE (%)	J_{sc} (mA/cm^2)	V_{oc} (mV)
—	YD2	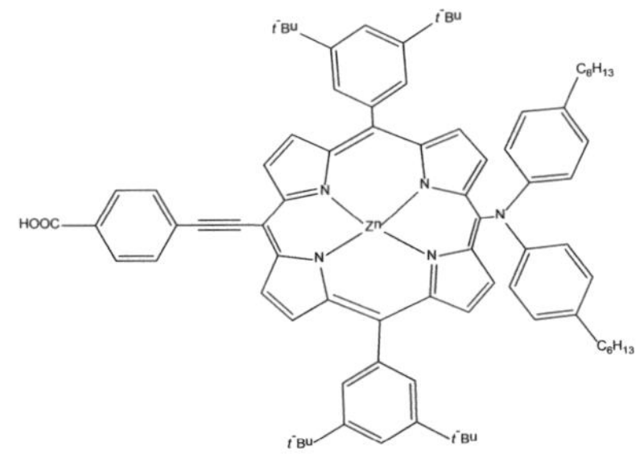	10.9	18.6	770
—	YD2-o-C8	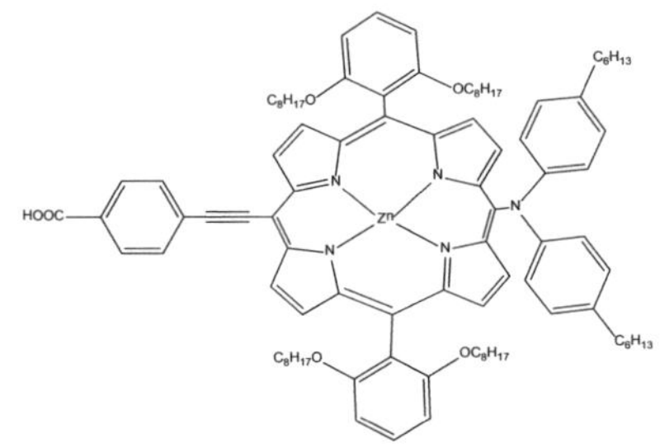	12.3	17.3	965
—	GY21	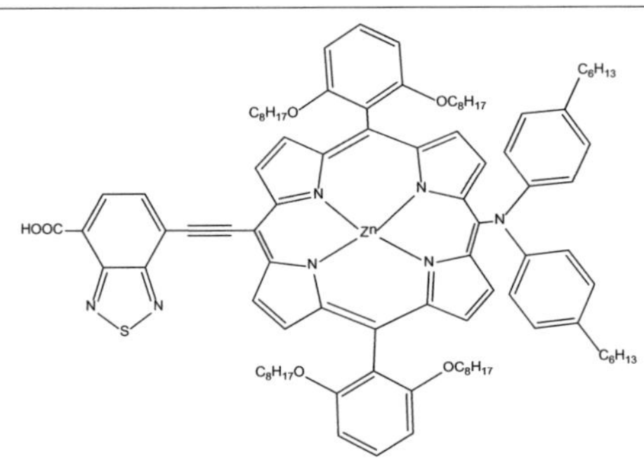	2.5	5.03	615
—	GY50	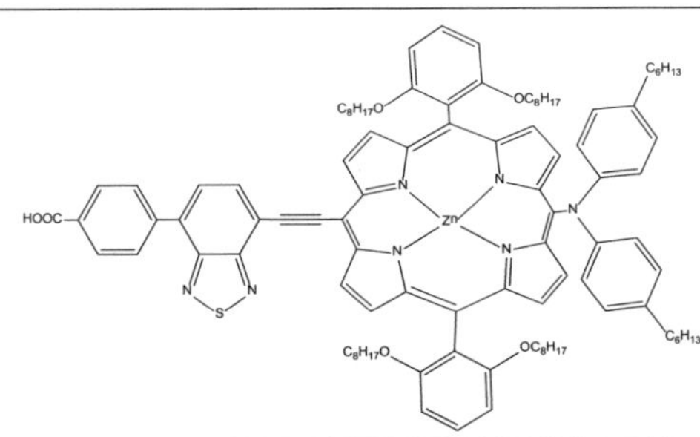	12.75	18.53	885
—	SM371	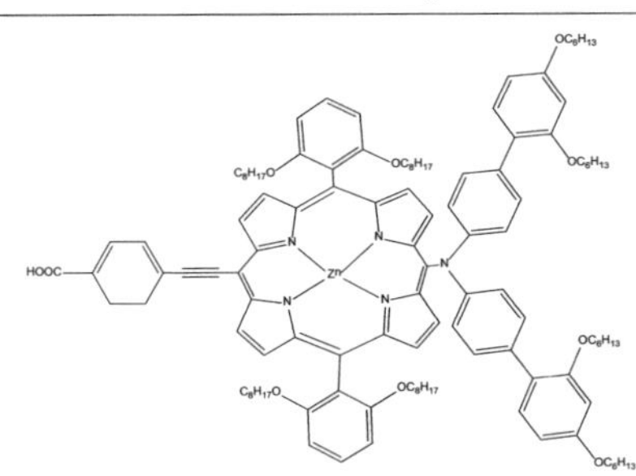	12	15.9	960

On the other hand, in this research also have been reported the theoretical studies towards the effect the spacer molecule in macrocycles. The linear molecule

Chemical name	Author's designation	Structure	PCE (%)	J_{sc} (mA/cm^2)	V_{oc} (mV)
—	SM 315	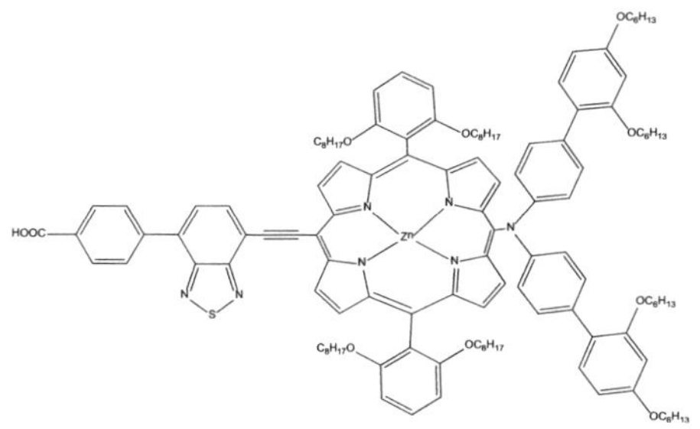	13	18.1	910
—	A4	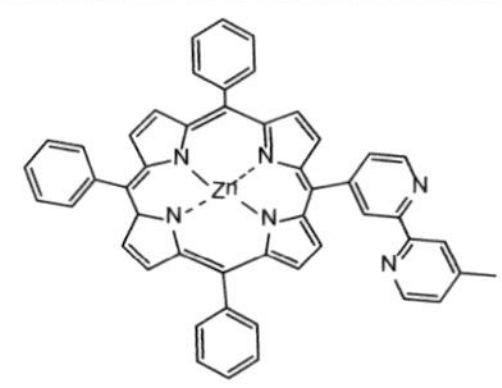	0.05	0.09	330
—	A6	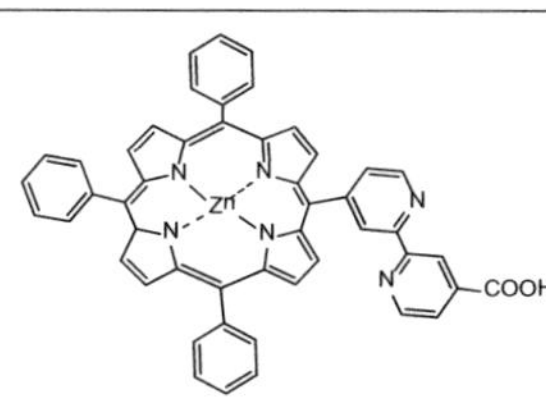	0.28	0.83	480
—	A7	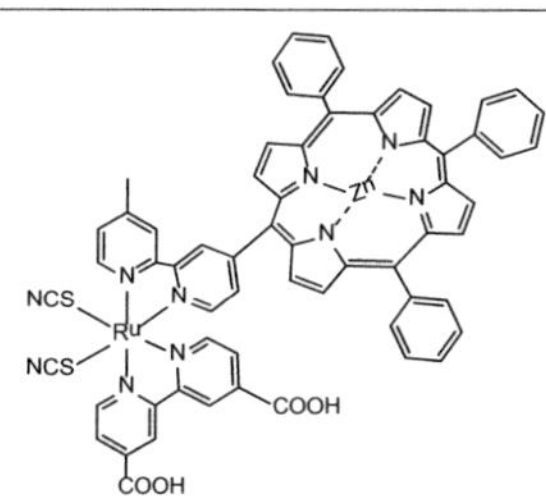	0.38	1.33	450
—	A8	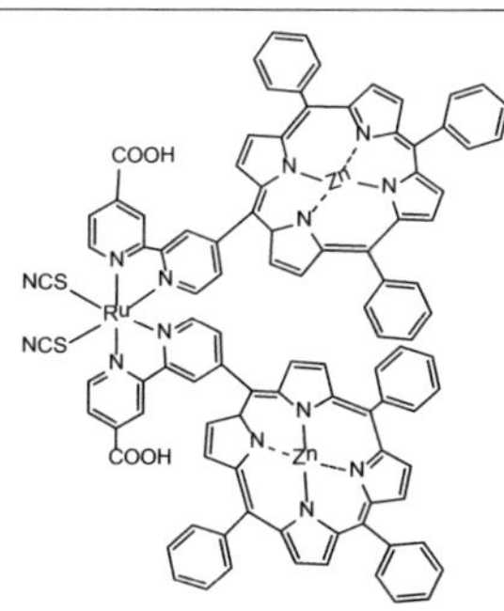	0.05	0.26	370
—	ZnT(Mes) P(CN-COOH)	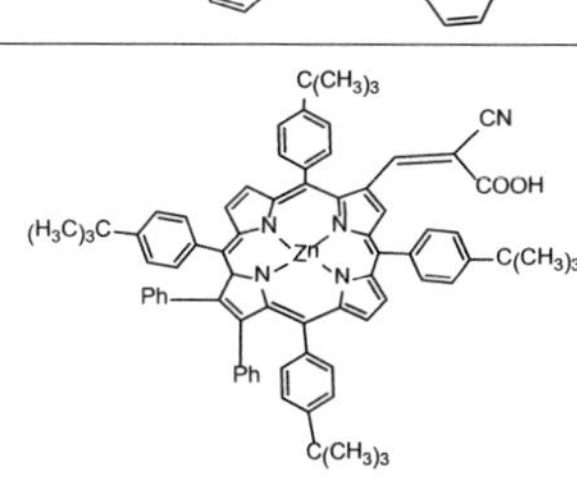	3.15	7.8	575

Chemical name	Author's designation	Structure	PCE (%)	J_{sc} (mA/cm^2)	V_{oc} (mV)
—	ZnT(4-t-Bu)P(Ph2)(CN-COOH)	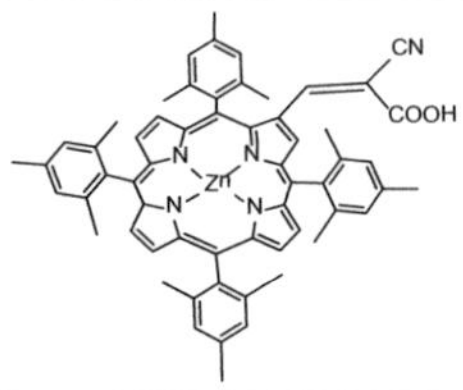	1.72	4.3	580
—	ZnPc1	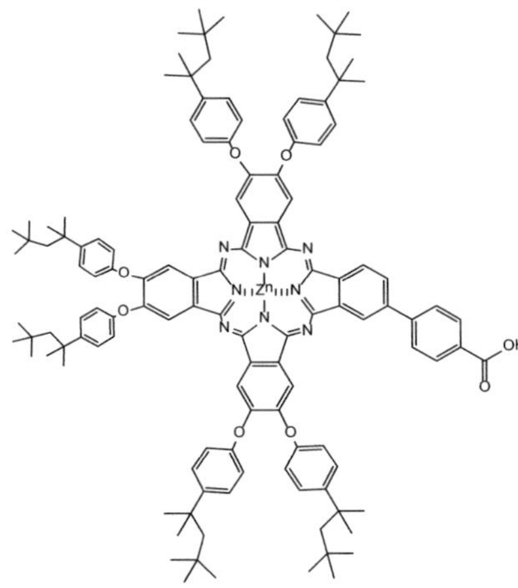	0.73	2.26	530
—	TT1	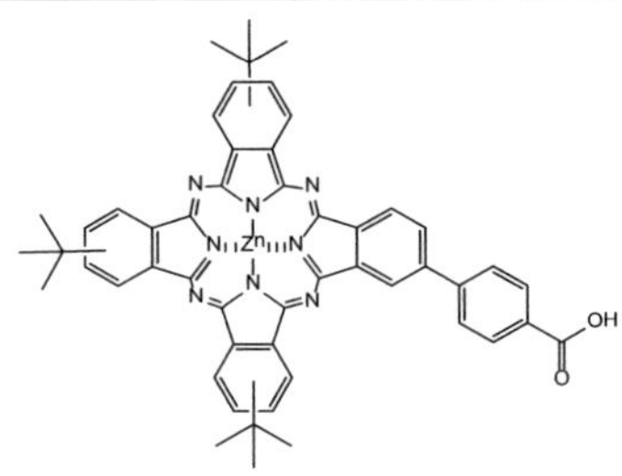	1.54	3.98	550

—, it is not mentioned in the article.

Table 2.
Porphyrins for DSSC [34–40].

designed was a benzothiophene derivate (T) and the spacer selected were *o- m* or *p-diphenyldiamine*. The spacer represented the communication channel between linear chains, denominated T. The stabilization of the macrocycles depends of the good assembly. The authors reported a study relationship with the photovoltaic properties for three macrocycles in function of isomeric effect in the spacer. The calculations were performed using Gaussian 09 16-18, program with B3LYP functional [58–61] and 6-31-6 (d, 2p) as basis set [64] in order to investigate the molecular geometry, electronic structures, and optical properties of *o*-PDT, *m*-PDT and *p*-PDT (**Figure 1**).

The stationary point was estimated with level of theory reported previously for the authors [64]. Finally, the authors through the Lewis acid incorporation showed an electronic improvement mechanism. The acid Lewis effect, as evaluated considering the tetracoordinated mode around metal center (**Figure 2**).

Chemical name	Author's designation	Structure	PCE (%)	J_{sc} (mA/cm^2)	V_{oc} (mV)	Electrolyte
2-Cyano-7-(1,1,6,6-tetramethyl-10-oxo-2,3,5,6-tetrahydro-1H,4H,10H-11-oxa-3a-aza-benzo[de]anthracen-9-yl)-hepta-2,4,6-trienoic acid	**NKX-2586**	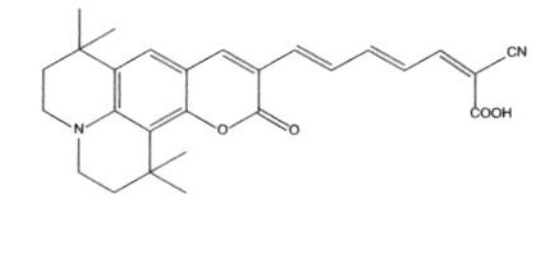	3.5	15.1	470	—
2-Cyano-5-(1,1,6,6-tetramethyl-10-oxo-2,3,5,6-tetrahydro-1H,4H,10H-oxa-3a-aza-benzo[de]anthracen-9-yl)-penta-2,4-dienoic acid	**NKX-2311**	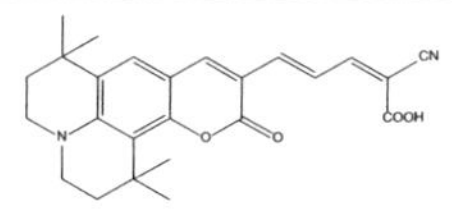	6.0	14.0	600	—
2-Cyano-3-[5′-(1,1,6,6-tetramethyl-10-oxo-2,3,5,6 tetrahydro-1H,4H,10H-11-oxa-3a azabenzo[de]anthracen-9-yl)-[2,2′]bithiophenyl-5-yl] acrylic acid	**NKX-2677**	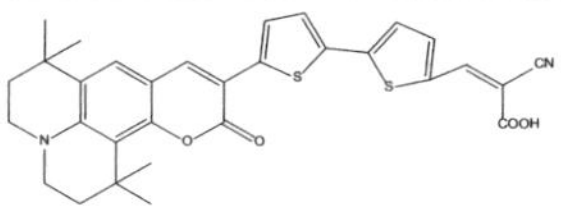	7.7	14.3	730	—
—	**D5**	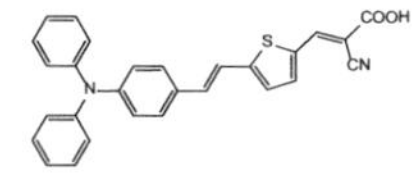	5.1	11.9	660	—
2-cyano-3-{5′-[1-cyano-2-(1,1,6,6-tetramethyl-10-oxo-2,3,5,6-tetrahydro-1H,4H,10H-11-oxa-3a-aza-benzo[de]anthracen-9-yl)-vinyl]-[2,2′]bithiophenyl-5-yl}-acrylic acid	**NKX-2883**	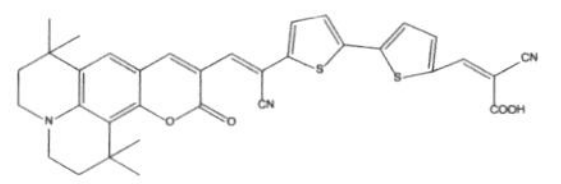	7.3	16.90	580	—
5-[[4-[4-(2,2-Diphenylethenyl)phenyl]-1,2,3,3a,4,8b-hexahydrocyclopent[*b*]indol-7-yl]methylene]-2-(3-octyl-4-oxo-2-thioxo-5-thiazolidinylidene)-4-oxo-3-thiazolidineacetic acid	**D205**	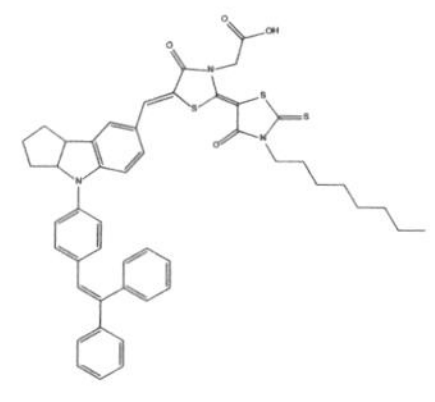	9.52	18.7	710	Ionic-liquid
2-Cyano-3-{5′-{2-{4-[N,N-bis(4-(2-ethylhexyloxy)-phenyl)amino]phenyl}-3,4-ethylenedi-oxythiophene-5-yl}-3,3′-di-n-hexylsilylene-2,2′-bithiophene-5-yl} acrylic acid	**C219**	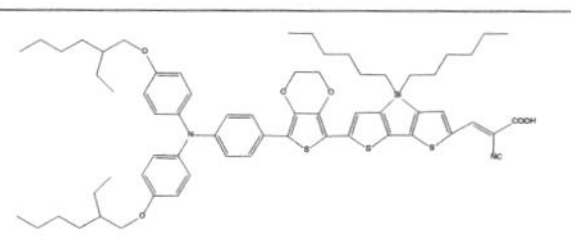	8.9	17.94	770	I_3/I^-
—	**C218**	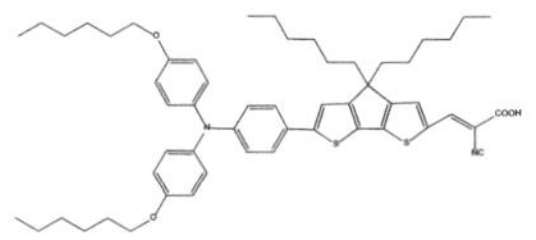	8.95	15.8	768	I_3/I^-

Chemical name	Author's designation	Structure	PCE (%)	J_{sc} (mA/cm^2)	V_{oc} (mV)	Electrolyte
2-cyano-3-[5′-(1,1,6,6-tetramethyl-10-oxo-2,3,5,6-tetrahydro-1H,4H,10H-11-oxa-3a-aza-benzo[de] anthracen-9-yl)-[2,1,3-benzothiadiazole]-4-thiophen-2-yl]-acrylic acid	**HKK-CM4**	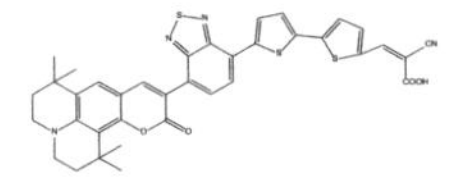	5.97	14.3	580	—
2-cyano-3-[5′-(1,1,6,6-tetramethyl-10-oxo-2,3,5,6-tetrahydro-1H,4H,10H-11-oxa-3a-aza-benzo[de] anthracen-9-yl)-[2,1,3-benzothiadiazole]-4-(3,4-ethylene dioxythiopheneyl-5-yl)]-acrylic acid	**HKK-CM5**	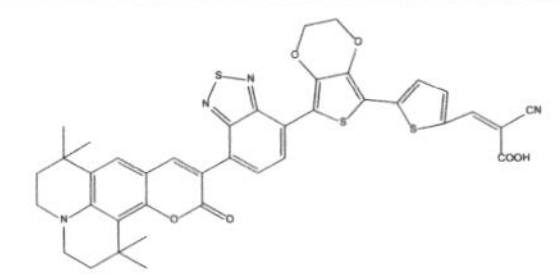	5.03	13.3	560	—
—	**C228**	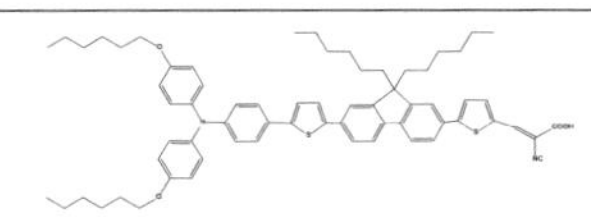	4.7	7.6	830	Co^{2+}/Co^{3+}
—	**C228**	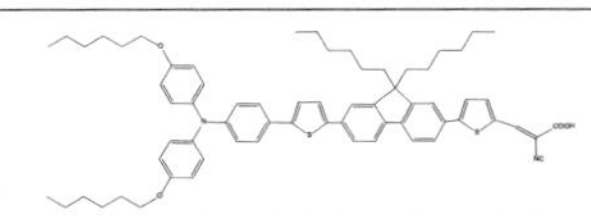	4.4	7.78	760	I_3/I^-
—	**C229**	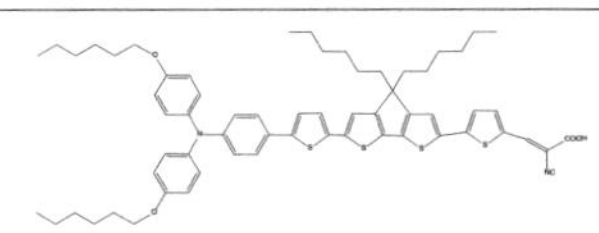	9.4	15.3	850	Co^{2+}/Co^{3+}
—	**C229**	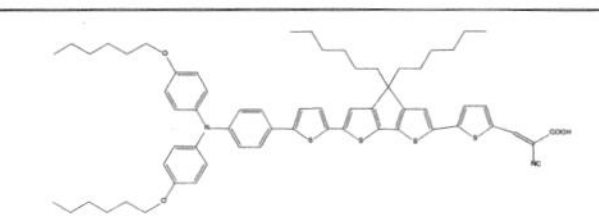	6,7	15.20	680	I_3/I^-
3-{6-{4-[bis (2′,4′ dihexyloxybiphenyl-4-yl)amino-]phenyl}-4,4-di-hexyl-cyclopenta-[2,1-b:3,4-b′]dithiphene-2-yl}-2-cyanoacrylic acid	**Y123**	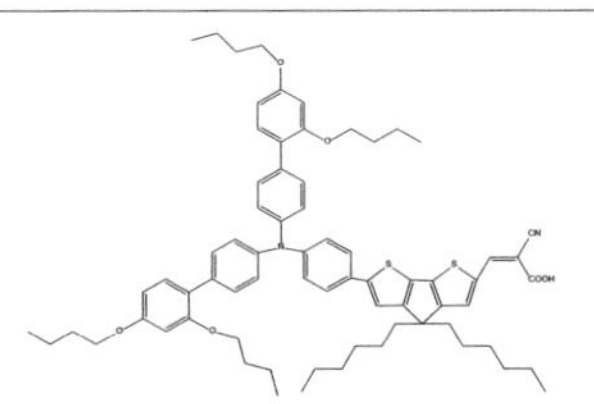	12.3	17.7	935	Co^{2+}/Co^{3+}
—	**ADEKA-1**	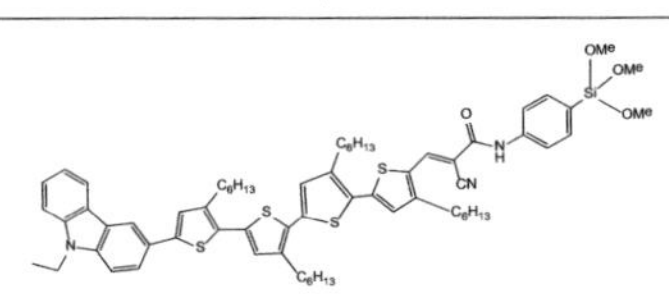	11.2	19.11	783	I_3/I^-
—	**LEG 4**	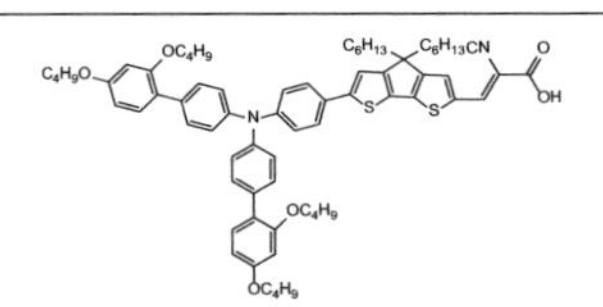	14.7	9.55	776	I_3/I^-

Chemical name	Author's designation	Structure	PCE (%)	J_{sc} (mA/cm^2)	V_{oc} (mV)	Electrolyte
—	**KNS-1**	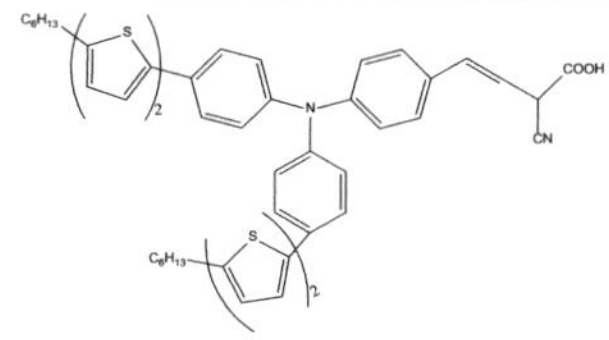	2.01	4.93	600	Co^{2+}/Co^{3+}
—	**KNS-2**	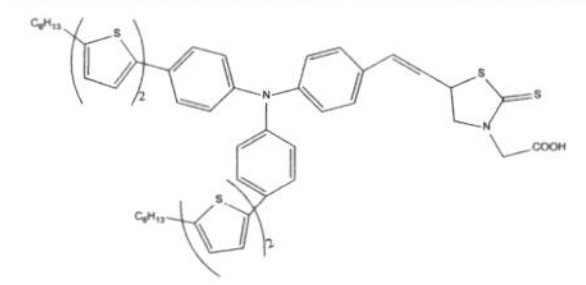	2.95	6.92	650	Co^{2+}/Co^{3+}
—	**JM-2**	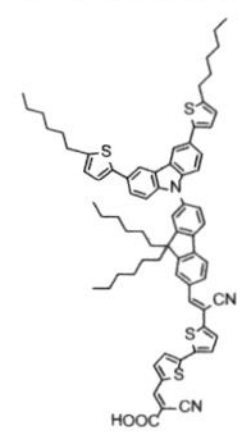	6.5	14.4	620	I_3/I^-
3-(5-((3,6-bis(bis(4-methoxyphenyl) amino)-9H-fluoren-9-ylidene)methyl) thiophen-2-yl)-2-cyanoacrylic acid	**TK-4**	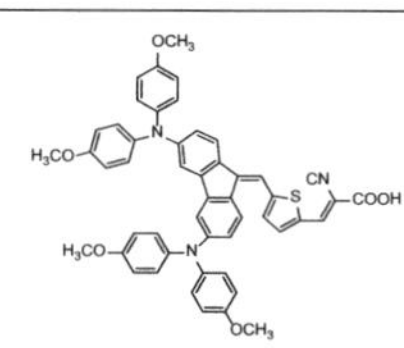	5.9	13.29	667	—
3-(5′-((3,6-bis(bis(4-methoxyphenyl) amino)-9H-fluoren-9-ylidene)methyl)-[2,2′-bithiophen]-5-yl)-2-cyanoacrylic acid	**TK-5**	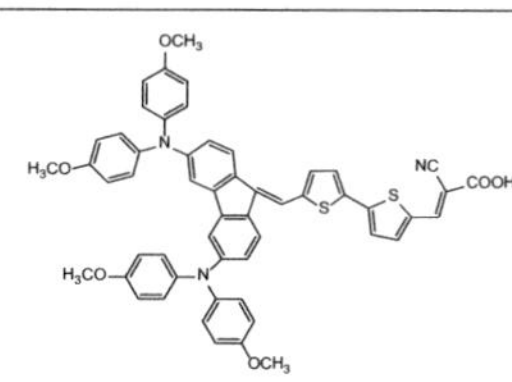	7.5	17.85	653	—
3-(5′-((3,6-bis(bis(4-(octyloxy)phenyl) amino)-9H-fluoren-9-ylidene)methyl)-2,2′-bithiophen-5-yl)-2-cyanoacrylic acid	**TK-6**	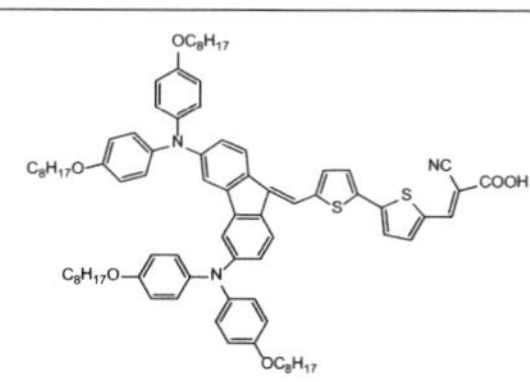	7.8	17.19	663	—
—	**O4T**	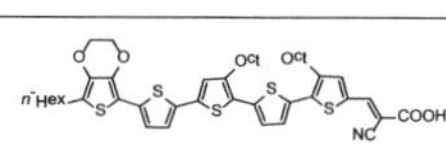	5.07	10.7	630	—
—	**ST4**	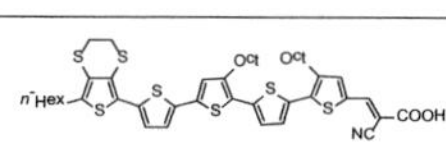	6.73	14.4	640	—
—	**P2**	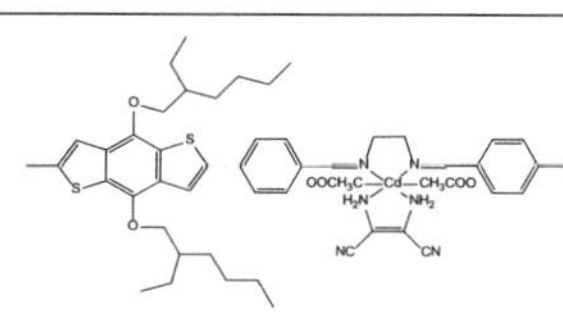	2.18	4.85	680	—
—	**LS-385**	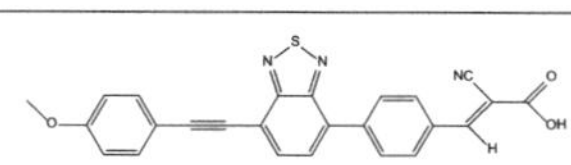	2.68	6.33	582	I_3/I^-
—	**LS-386**	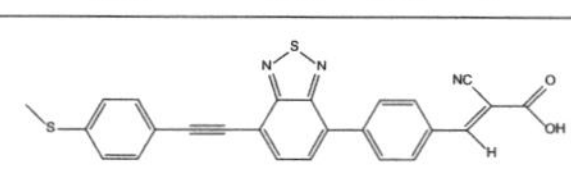	2.69	6.53	561	I_3/I^-

Chemical name	Author's designation	Structure	PCE (%)	J_{sc} (mA/cm^2)	V_{oc} (mV)	Electrolyte
—	LS-387	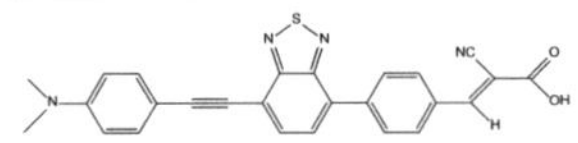	5.61	13.26	595	I_3/I^-
—	TA1	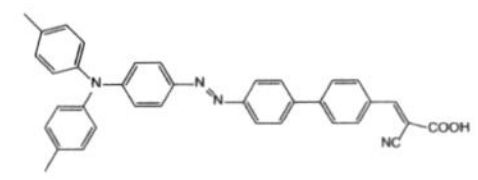	2.56	5.40	662	—
—	TA2	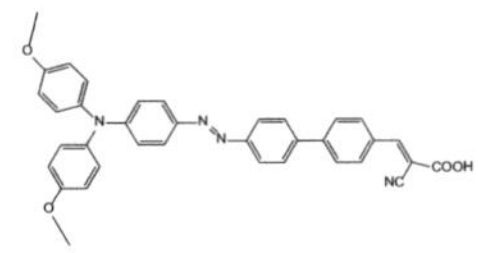	3.45	6.83	704	—
—	TA3	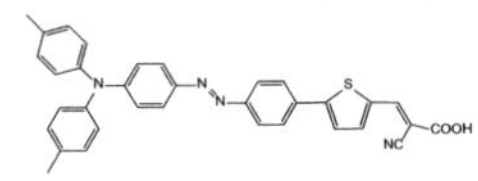	3.69	7.81	654	—
—	TA4	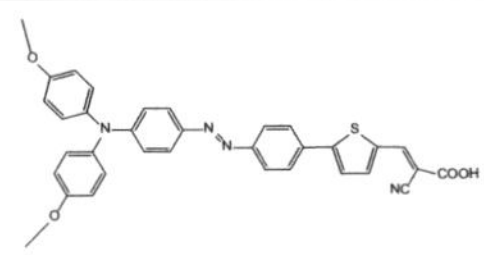	4.78	9.92	662	—
(Z)-2-cyano-3-(5″-((E)-2,4,6-trimethoxystyryl)-[2,2′:5′,2″-terthiophen]-5-yl) acrylic acid	MR-5	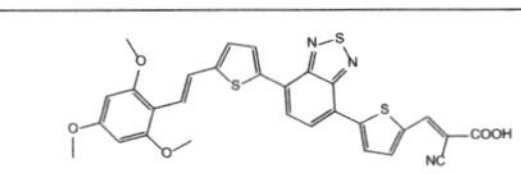	6.03	15.27	610	—
2-((Z)-4-oxo-2-thioxo-5-((5″-((E)-2,4,6-trimethoxystyryl)-[2,2′:5′,2″-terthiophen]-5-yl) methylene) thiazolidin-3-yl) acetic acid	MR-6	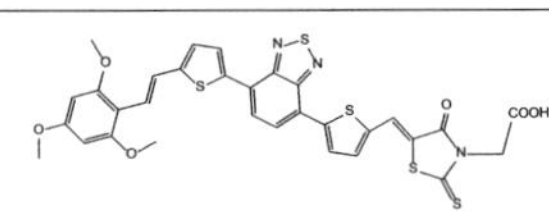	3.2	8.7	560	—
—	D6	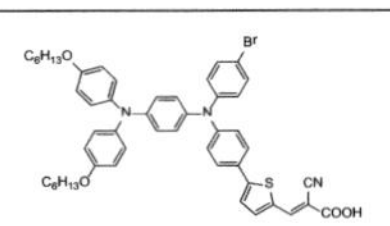	4.7	8.6	793	Co^{2+}/Co^{3+}
—	D7	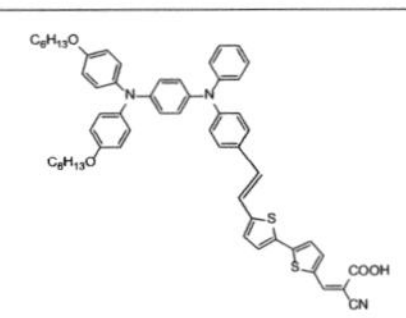	4.0	8.2	768	Co^{2+}/Co^{3+}
—	SP1	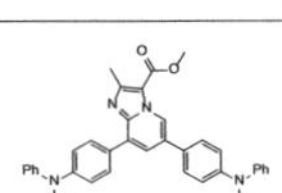	0.86	2.59	625	—
—	SP3	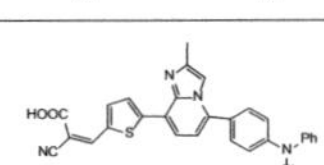	0.43	2.31	532	—
—	SP4	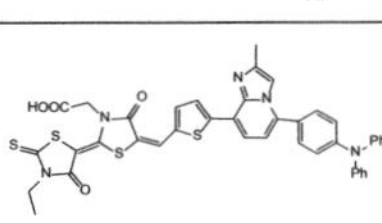	0.58	0.78	290	—
—	BD-3	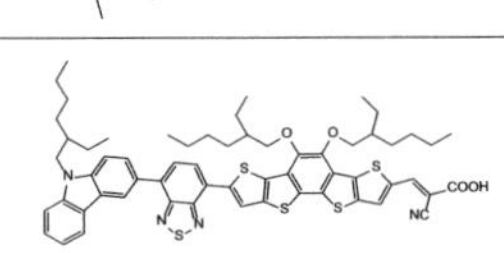	5.46	12.21	680	I_3/I^-

Chemical name	Author's designation	Structure	PCE (%)	J_{sc} (mA/cm^2)	V_{oc} (mV)	Electrolyte
—	BD-5	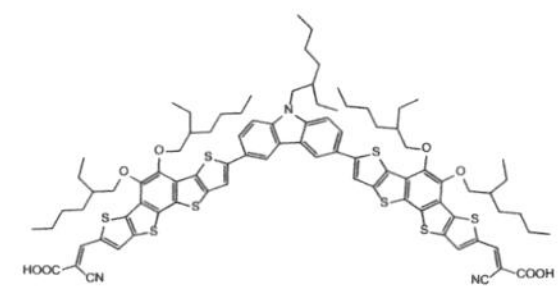	5.34	12.23	680	I_3/I^-
—	Dendrimer 1	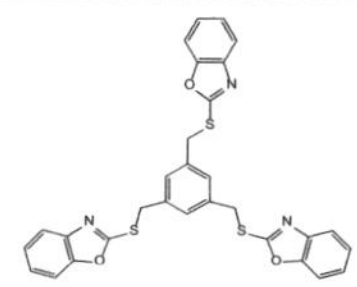	5.19	13.00	675	I_3/I^-
—	4a	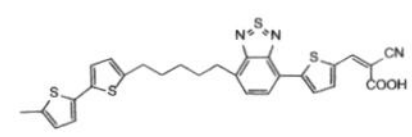	0.95	2.60	500	—
—	4b	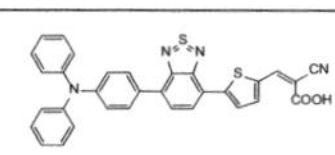	4.51	9.36	630	—
—	N, N′-PABA	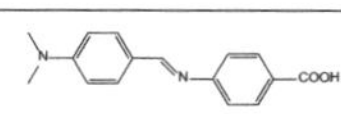	1.00	2.72	537	I_3/I^-

—, it is not mentioned in the article.

Table 3.
Structures associated with metal-free organic dyes [41–63].

2. Material and methods

2.1 Method theoretical

To explore and understand the electronic and optoelectronic properties of photosensitized materials with application in OPV technology, many theories have emerged. One of the most important and common theories is the theory of functional density (DFT), which is a tool that allowed to establish any property used in photosensitized materials, quantum state of atoms, molecules and solids, making modeling and simulation possible of complex systems with millions of degrees of freedom. At present, DFT has grown tremendously and has become one of the main tools in theoretical physics and molecular chemistry. Modeling in the framework of computational chemistry of photosensitized systems made up of electron donors and electron acceptors ultimately influences photo induced electron transfer and energy reactions. Numerous studies using the Density Functional Theory (DFT) methodology to design, evaluate and predict photovoltaic properties of photoactive materials with application in OPV have been published. The approximation of the theory of functional density (DFT) implemented was Gaussian 09 together with the functional correlation (B3LYP) and the base set 6-31g (d, 2p). This calculation allows optimization of geometry without symmetry restrictions for stationary points. In addition, it provided information on the harmonic frequency analysis, which allows the optimized minimum to be verified. The local minimum is identified when the number of imaginary 32frequencies is equal to zero.

The analysis of the changes in electron density for a given electronic transition was based on the electron density difference maps (EDDMs) constructed using the GaussSum suite of programs. Gásquez and co-workers had proposed two different electronegativities (X) for the charge transfer process: one that describes fractional

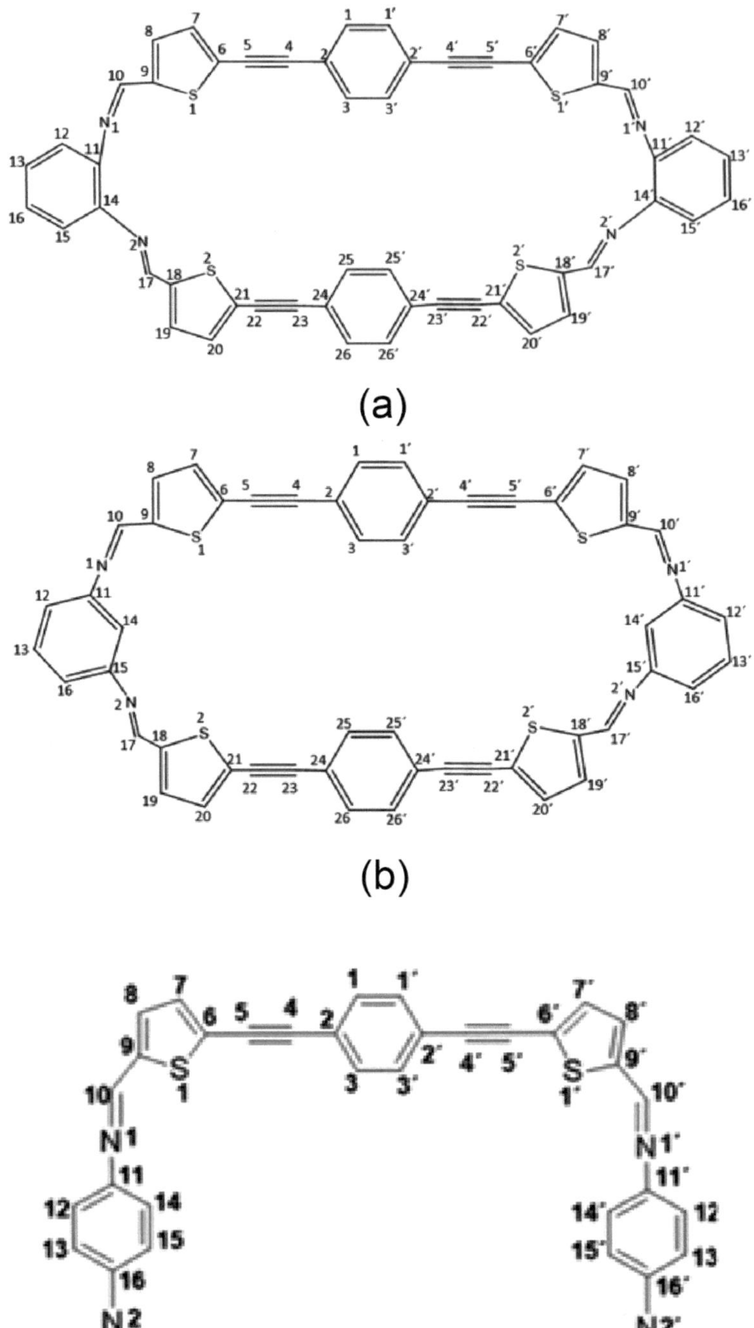

Figure 1.
Chemical structure for (a) p-*PDT, (b)* m-*PDT and (c)* o-*PDT.*

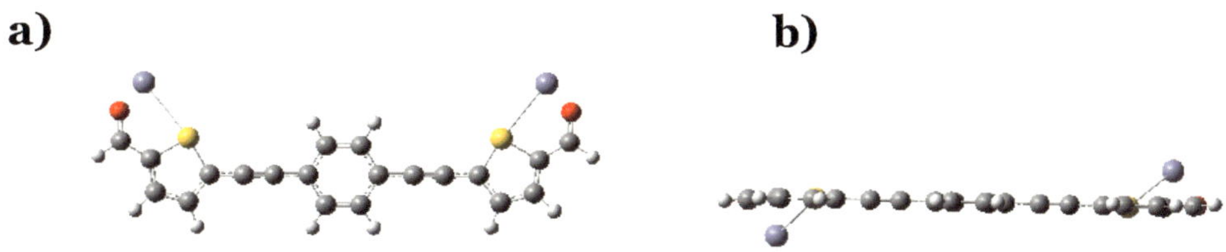

Figure 2.
Geometry optimization for (a) T and (b) TZn.

negative charge donation X^- whereas the other gives the fractional negative charge acceptance X^+, Eqs. (1) and (2):

$$X^- = 0.25(3I + A) \quad (1)$$

$$X^+ = 0.25\,(I + 3A) \quad (2)$$

Thus, the construction of a so-called donor-acceptor map (DAM) has been suggested. A DAM graphic can be constructed by plotting the values of (y-axis) and X^+ (x-axis) for each molecule of interest.

The photovoltaic properties are calculated according to the Scharber model, which is an empirical model for predicting the PCE of organic cell solar. HOMO-LUMO as orbital border under solar irradiation with AM 1.5 G (ASTM G173). The PCE was expressed by the following Eqs. (3) and (4), in where V_{oc} is the open circuit voltage, and J_{sc} is short circuit current.

$$\boldsymbol{PCE} = FF\left(J_{sc}\,\frac{V_{oc}}{P_{inc}}\right) \quad (3)$$

where *FF* is a fill factor of 0.75, Eq. (4):

$$\mathbf{J_{sc}} = \mathrm{q}\int_{\lambda min}^{\lambda max} EQE\ \phi^{AM\ 1.5\ G}(\lambda)d\lambda \quad (4)$$

(q = elementary charge, EQE = external quantum efficiency, ϕ = irradiation flow with AM 1.5 G, and λ = wavelength), and P_{inc} (incident light power).

On the other hand, the valor corresponding ΔE_{GAP} was calculated as, Eq. (5):

$$\Delta E = E\ (LUMO) - E\ (HOMO) \quad (5)$$

LHE (light capture efficiency determinate), (f = oscillator strength) and E_{s1} = Excitation energy for λ_{max}, Eq. (6).

$$LHE = 1 - 10^{-f} \quad (6)$$

3. Results and discussion

3.1 Optical properties of macrocycle molecules

The excitation energies (Es1) presented in **Table 4** were relatively small for the PDT molecule ligand, which indicate a shift to visible region in relationship with λ_{max}. The *p*-PDT showed the lowest value for E_{s1}, which is directly correlated with the conversion energy (PCE). On the other hand, the cycle size generated for *o*-PDT

Molecule	Wavelength λ_{max}	E_{s1} (eV)	*F*	LHE
o-PDT	432.68	2.874	2.0217	0.9905
m-PDT	425.37	2.923	2.8755	0.9987
p-PDT	465	2.67	0.07	0.14

Table 4.
Optical properties for (a) o-PDT, *(b)* m-PDT *and (c)* p-PDT.

and *m*-PDT systems is smaller, but this does not guarantee a better transfer. On the contrary, there is less efficient in the electronic transport.

The LHE values were 2.0217, 2.8755 and 0.07 for *o*-PDT, *m*-PDT and *p*-PDT, respectively. This indicates that *o*-PDT and *m*-PDT had a similar sensitivity to sunlight and will reflect higher values of LHE compared to *p*-PDT.

The visible light absorption ability may benefit from absorbing more photons and generating high photocurrent, which is a strong advantage of T derivatives. In the previous reports, PD spacers that cannot absorb visible light were observed. It is necessary that T derivate linked to the PD fragment enhances the electronic coupling in the excited state, which operates as a gated wire in π-conjugated systems, as is observed for *o*-PDT, *m*-PDT and *p*-PDT (**Figure 3**). The isomeric effect is greatly correlated to geometric distortion *o*-PDT and *m*-PDT molecules, which were dramatically affected in relationship to its planarity. The cavity between linear molecules is small, but the torsion affects the electronic properties.

3.2 Geometry study for macrocycles with Lewis acid.

The effect of Lewis acid on macrocycle stabilization is shown below. The geometric environment of the metallic center was tetrahedral, considering two positions to the electro donator atoms corresponding to linear and macrocycle molecule; and two water molecules. The incorporation of the metal into the linear chain (ZnT) generates a decrease in the value for GAP around 1.72 eV, a value located in the visible region. However, the effect is more severe when incorporated into the macrocycle, in where; its addition generated a decrease in GAP still 1.55 eV (*p*-ZnPDT). The DAM graphic for these systems indicated a significant improvement in donor capacity. These criteria are important to electronically activate the photovoltaic cell (**Figure 4**).

The spectra in **Figure 5** showed a similar profile for TZn, and *p*-ZnPDT with the incorporation of Lewis acid in the structure, which have an intense main band to 568 nm, and 516 nm respectively. This band corresponds to the dominant electron transition from HOMO to LUMO, that is, from the π molecular orbital (chromophore fragments-π-linker) to the π* orbital (acceptor fragment), and this process can be ascribed to the intramolecular charge transfer.

3.3 Photovoltaic properties of macrocycle molecules

The results showed in **Table 5** suggested decreased the ΔE_{GAP} in relationship with PCE. These results are congruent with the optical, and electronic properties observed previously. The *p*-PDT presented the best photovoltaic properties. The metal ion generates a symmetrical tension in the system, and this could explain its behavior. The J_{sc} increased in function of decreased the ΔE_{GAP}, concluding that the preferential isomer for the construction of this family macrocycles is the *p*-PDT, considering theoretical models in the gas phase.

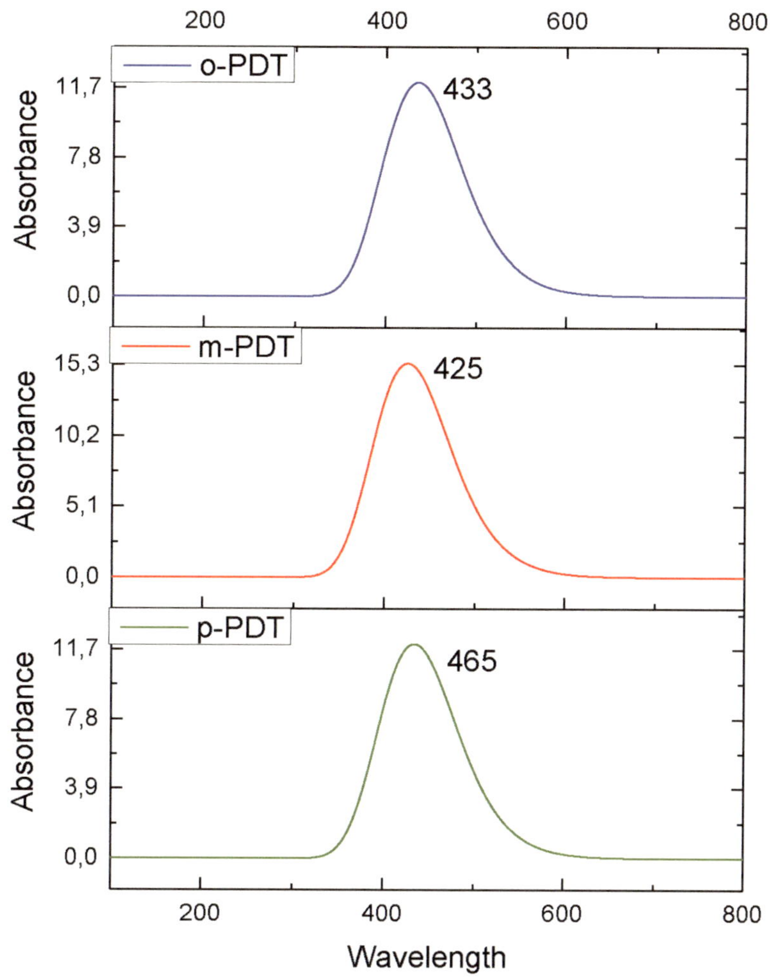

Figure 3.
Theoretical spectra electronic for (a) o-*PDT* m-*PDT.*

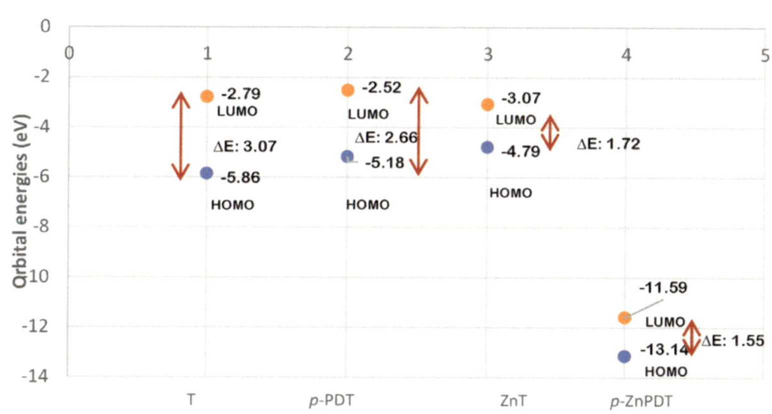

Figure 4.
DAM graphic for T, p-*PDT, ZnT and* p-*ZnPDT.*

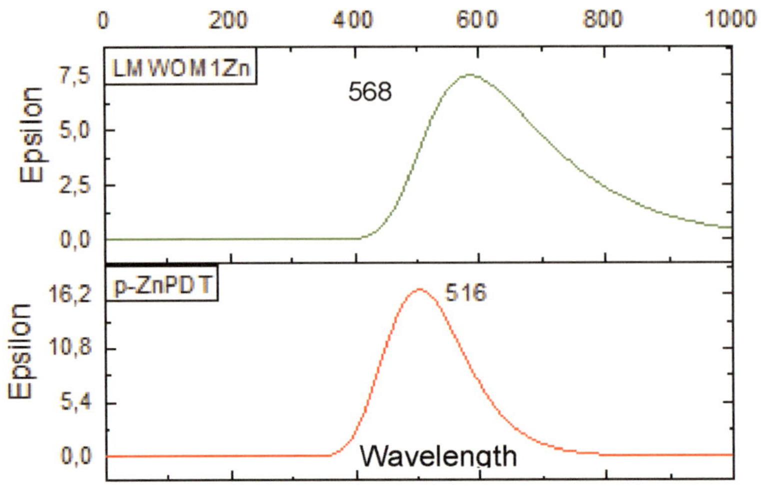

Figure 5.
Theoretical spectra electronic for TZn and p-*ZnPDT.*

Molecule	ΔE_{GAP} (eV)	J_{sc} (mA/cm^2)	V_{oc} (V)	FF	PCE (%)
T	3.07	5.36	2.77	0.75	11.12
p-PDT	2.66	14.79	2.36	0.75	26.18
ZnT	1.72	21.52	1.42	0.75	22.92
p-ZnPDT	1.55	11.76	1.25	0.75	11.02

Table 5.
Photovoltaic parameters for T, p-*PDT, ZnT and* p-*ZnPDT.*

4. Summary and future perspectives

The purpose of this review of DSSC materials was to compile the information reported to: synthetized ruthenium complexes, porphyrins, and metal-free organic dyes. For researchers, it is important to know parameters such as: PCE, J_{sc}, and V_{oc}; which help you to diffuse between structures, and propose synthesis strategies that make possible new materials in this field application. Principles for the future development of new molecules can be analyzed and likewise it is interesting support to follow up structure families as a function of time. Although many structures are shown here, there is still a need to optimize the chemical, and physical properties to promote improved solar cells. On the other hand, in this work, the best photovoltaic parameters were described for *p*-PDT with PCE 26.18%, J_{sc} = 14.79 mA/cm^2, and ΔE = 2.66 eV such as macrocycle. The metal ion influences the electronic properties, and decreases the ΔE_{GAP}. The incorporation of Lewis acid in the structure macrocycle to increase of the optical properties, which allows rigidity that can benefit planarity.

Acknowledgements

This work was supported by Universidad Santiago de Cali—DGI Grants 63661. The author acknowledgment to Melissa Suarez for technical support and Hoover Valencia for data acquisition.

Conflict of interest

The authors declare that they have no conflict of interests.

Author details

Yenny Patricia Avila Torres
Research Group QUIBIO, Facultad de Ciencias Básicas, Universidad Santiago de Cali, Santiago de Cali, Colombia

*Address all correspondence to: yavilatorres@gmail.com

References

[1] Owusu PA, Asumadu-Sarkodie S. A review of renewable energy sources, sustainability issues, and climate change mitigation. Cogent Engineering. 2016; **3**(1):1167990

[2] Wang P, Zakeeruddin SM, Moser JE, Humphry-Baker R, Comte P, Aranyos V, et al. Stable new sensitizer with improved light harvesting for nanocrystalline dye-sensitized solar cells. Advanced Materials. 2004;**16**(20): 1806-1811

[3] Hara K, Sugihara H, Tachibana Y, Islam A, Yanagida M, Sayama K, et al. Dye-sensitized nanocrystalline TiO_2 solar cells based on ruthenium (II) phenanthroline complex photosensitizers. Langmuir. 2001;**17**(19):5992-5999

[4] Manivel P, Mohana Roopan S, Nawaz Khan F. Synthesis of O-substituted benzophenones by Grignard reaction of 3-substituted isocoumarins. Journal of the Chilean Chemical Society. 2008; **53**(3):1609-1610

[5] Philipps SP, Bett AW, Horowitz K, Kurtz S. Current Status of Concentrator Photovoltaic (CPV) Technology. No. NREL/TP-5J00-65130. Golden, CO (United States): National Renewable Energy Lab. (NREL); 2015

[6] Dusastre V. Materials for sustainable energy: A collection of peer-reviewed research, and review articles. Editorial. Nature Publishing Group, World Scientific; 2011

[7] O'regan B, Grätzel M. A low-cost, high-efficiency solar cell based on dye-sensitized colloidal TiO_2 films. Nature. 1991;**353**(6346):737

[8] Mathews L, Tidwell C, Bharara P, Stephens G, Su T, Carter A. Copper 5, 10, 15, 20-Tetrakis-(3, 4-dibenzyloxyphenyl) porphyrin. Molbank. 2017;**1**(2017):M931

[9] Robertson N. Optimizing dyes for dye-sensitized solar cells. Angewandte Chemie International Edition. 2006; **45**(15):2338-2345

[10] Xie P, Guo F. Molecular engineering of ruthenium sensitizers in dye-sensitized solar cells. Current Organic Chemistry. 2007;**11**(14):1272-1286

[11] Bessho T, Zakeeruddin SM, Yeh C-Y, Diau EW-G, Grätzel M. Highly efficient mesoscopic dye-sensitized solar cells based on donor–acceptor-substituted porphyrins. Angewandte Chemie International Edition. 2010; **49**(37):6646-6649

[12] Schiffmann F, VandeVondele J, Hutter J, Urakawa A, Wirz R, Baiker A. An atomistic picture of the regeneration process in dye sensitized solar cells. Proceedings of the National Academy of Sciences. 2010;**107**(11):4830-4833

[13] Nazeeruddin MK, Kay A, Rodicio I, Humphry-Baker R, Müller E, Liska P, et al. Conversion of light to electricity by cis-X2bis (2, 2′-bipyridyl-4, 4′-dicarboxylate) ruthenium (II) charge-transfer sensitizers (X= Cl-, Br-, I-, CN-, and SCN-) on nanocrystalline titanium dioxide electrodes. Journal of the American Chemical Society. 1993; **115**(14):6382-6390

[14] Yu Q, Wang Y, Yi Z, Ningning Z, Zhang J, Zhang M, et al. High-efficiency dye-sensitized solar cells: The influence of lithium ions on exciton dissociation, charge recombination, and surface states. ACS Nano. 2010;**4**(10): 6032-6038

[15] Duan C, Huang F, Cao Y. Recent development of push–pull conjugated polymers for bulk-heterojunction photovoltaics: Rational design and fine tailoring of molecular structures. Journal of Materials Chemistry. 2012;**22**(21): 10416-10434

[16] Chen G, Sasabe H, Igarashi T, Hong Z, Kido J. Squaraine dyes for organic photovoltaic cells. Journal of Materials Chemistry A. 2015;**3**(28): 14517-14534

[17] Xunjin ZHU, Chen S, Wong WY, Wong WK. Design and synthesis of porphyrin materials for highly efficient organic photovoltaics. U.S. Patent Application 15/635,563. 2018

[18] Suarez M, Patricia Avila-Torres Y, Caicedo C, Guateque J, Florez-Lopez E. Novel macrocyclic thiazole as potential organic photovoltaic. In: Proceedings of the 10 th European meeting on Solar Chemistry and Photocatalysis: Environmental Applications (SPEA10). 2018. pp. 626-628

[19] Frisch MJEA, Trucks GW, Schlegel HB, Scuseria GE, Robb MA, Cheeseman JR, et al. Gaussian 09, Revision a. 02. Wallingford, CT: Gaussian. Inc.; 2009. p. 200

[20] Hohenberg P, Kohn W. Inhomogeneous electron gas. Physical Review. 1964;**136**:B864

[21] Kohn W, Sham LJ. Self-consistent equations including exchange and correlation effects. Physical Review. 1965;**140**:A1133

[22] Astm G. 173-03: Standard Tables for Reference Solar Spectral Irradiances: Direct Normal and Hemispherical on 37 Tilted Surface. West Conshohoken, PA: ASTM International; 2003

[23] Lee C, Yang W, Parr RG. Development of the Colle-Salvetti correlation-energy formula into a functional of the electron density. Physical Review B. 1988;**37**:785

[24] Williams RM, De Cola L, Hartl F, Lagref JJ, Planeix JM, De Cian A, et al. Photophysical, electrochemical and electrochromic properties of copper-bis (4, 4′-dimethyl-(32)6, 6′-diphenyl-2, 2′-bipyridine) complexes. Coordination Chemistry Reviews. 2002;**230**:253-261

[25] Lee S, Seok CH, Park Y, Lee A, Jung DH, Choi SH, et al. Enforced effects of side group substitution position on luminescence properties; synthesis of Bis (dipyrrinato) zinc complex derivatives. Molecular Crystals and Liquid Crystals. 2010;**531**: 65-365

[26] Fitri A, Benjelloun AT, Benzakour M, Mcharfi M, Hamidi M, Bouachrine M. Theoretical design of thiazolothiazole-based organic dyes with different electron donors for dye-sensitized solar cells. Spectrochimica Acta Part A: Molecular and Biomolecular Spectroscopy. 2014;**132**: 232-238

[27] Nazeeruddin MK, Pechy P, Renouard T, Zakeeruddin SM, Humphry-Baker R, Comte P, et al. Engineering of efficient panchromatic sensitizers for nanocrystalline TiO_2-based solar cells. Journal of the American Chemical Society. 2001; **123**(8):1613-1624

[28] Gao F, Wang Y, Dong S, Zhang J, Wang M, Jing X, et al. Enhance the optical absorptivity of nanocrystalline TiO_2 film with high molar extinction coefficient ruthenium sensitizers for high performance dye-sensitized solar cells. Journal of the American Chemical Society. 2008;**130**(32):10720-10728

[29] Chen C-Y, Wang M, Li J-Y, Pootrakulchote N, Alibabaei L, Ngoc-le C-h, et al. Highly efficient light-harvesting ruthenium sensitizer for thin-film dye-sensitized solar cells. ACS Nano. 2009;**3**(10):3103-3109

[30] Swetha T, Niveditha S, Bhanuprakash K, Singh SP. Panchromatic Ru (II) Dipyrrins as NCS free sensitizers showing highest efficiency for DSSCs. Electrochimica Acta. 2015;**153**:343-351

[31] Naik P, Elmorsy MR, Rui S, Babu DD, El-Shafei A, Adhikari AV. New carbazole based metal-free organic dyes with D-π-A-π-a architecture for DSSCs: Synthesis, theoretical, and cell performance studies. Solar Energy. 2017;**153**:600-610

[32] Dayan S, Özpozan NK. Performance improvement of RuII complexes pyridinyl backbone on dye-sensitized solar cells (DSSC). Inorganica Chimica Acta. 2018;**474**:81-88

[33] Lyu S, Bertrand C, Hamamura T, Ducasse L, Toupance T, Olivier C. Molecular engineering of ruthenium-diacetylide organometallic complexes towards efficient green dye for DSSC. Dyes and Pigments. 2018;**158**:326-333

[34] Hsieh C-P, Lu H-P, Chiu C-L, Lee C-W, Chuang S-H, Mai C-L, et al. Synthesis and characterization of porphyrin sensitizers with various electron-donating substituents for highly efficient dye-sensitized solar cells. Journal of Materials Chemistry. 2010;**20**(6):1127-1134

[35] Yella A, Lee H-W, Tsao HN, Yi C, Chandiran AK, Nazeeruddin MK, et al. Porphyrin-sensitized solar cells with cobalt (II/III)–based redox electrolyte exceed 12 percent efficiency. Science. 2011;**334**(6056):629-634

[36] Yella A, Mai C-L, Zakeeruddin SM, Chang S-N, Hsieh C-H, Yeh C-Y, et al. Molecular engineering of push–pull porphyrin dyes for highly efficient dye-sensitized solar cells: The role of benzene spacers. Angewandte Chemie International Edition. 2014;**53**(11): 2973-2977

[37] Mathew S, Yella A, Gao P, Humphry-Baker R, Curchod BFE, Ashari-Astani N, et al. Dye-sensitized solar cells with 13% efficiency achieved through the molecular engineering of porphyrin sensitizers. Nature Chemistry. 2014;**6**(3):242

[38] Duanglaor P, Thiampanya P, Sudyoadsuk T, Promarak V, Pulpoka B. Synthesis and photophysical properties of donor–acceptor system based bipyridylporphyrins for dye-sensitized solar cells. Journal of Energy Chemistry. 2015;**24**(6):779-785

[39] Prakash K, Manchanda S, Sudhakar V, Sharma N, Sankar M, Krishnamoorthy K. Facile synthesis of β-functionalized "push-pull" Zn (II) porphyrins for DSSC applications. Dyes and Pigments. 2017;**147**:56-66

[40] Martín-Gomis L, Parejo C, Álvarez JC, Fernández-Lázaro F, Sastre-Santos Á. Dye sensitized solar cells (DSSCs) based on bulky tert-octylphenoxy-carboxyphenyl substituted phthalocyanine without the presence of co-adsorbents. Inorganica Chimica Acta. 2017;**468**:327-333

[41] Hara K, Tachibana Y, Ohga Y, Shinpo A, Suga S, Sayama K, et al. Dye-sensitized nanocrystalline TiO_2 solar cells based on novel coumarin dyes. Solar Energy Materials and Solar Cells. 2003;**77**(1):89-103

[42] Hara K, Kurashige M, Dan-oh Y, Kasada C, Shinpo A, Suga S, et al. Design of new coumarin dyes having thiophene moieties for highly efficient organic-dye-sensitized solar cells. New Journal of Chemistry. 2003;**27**(5):783-785

[43] Hagberg DP, Edvinsson T, Marinado T, Boschloo G, Hagfeldt A, Sun L. A novel organic chromophore for dye-sensitized nanostructured solar cells. Chemical Communications. 2006; **21**:2245-2247

[44] Wang Z-S, Cui Y, Dan-oh Y, Kasada C, Shinpo A, Hara K. Thiophene-functionalized coumarin dye for efficient dye-sensitized solar cells: Electron lifetime improved by coadsorption of deoxycholic acid. The Journal of Physical Chemistry C. 2007; **111**(19):7224-7230

[45] Ito S, Miura H, Uchida S, Takata M, Sumioka K, Liska P, et al. High-conversion-efficiency organic dye-sensitized solar cells with a novel indoline dye. Chemical Communications. 2008;**41**:5194-5196

[46] Zeng W, Cao Y, Yu B, Wang Y, Shi Y, Zhang M, et al. Efficient dye-sensitized solar cells with an organic photosensitizer featuring orderly conjugated ethylenedioxythiophene and dithienosilole blocks. Chemistry of Materials. 2010;**22**(5):1915-1925

[47] Bai Y, Zhang J, Zhou D, Wang Y, Zhang M, Wang P. Engineering organic sensitizers for iodine-free dye-sensitized solar cells: Red-shifted current response concomitant with attenuated charge recombination. Journal of the American Chemical Society. 2011;**133**(30): 11442-11445

[48] Tsao HN, Yi C, Moehl T, Yum J-H, Zakeeruddin SM, Nazeeruddin MK, et al. Cyclopentadithiophene bridged donor–acceptor dyes achieve high power conversion efficiencies in dye-sensitized solar cells based on the tris-cobalt bipyridine redox couple. ChemSusChem. 2011;**4**(5):591-594

[49] Kakiage K, Aoyama Y, Yano T, Oya K, Fujisawa J-i, Hanaya M. Highly-efficient dye-sensitized solar cells with collaborative sensitization by silyl-anchor and carboxy-anchor dyes. Chemical Communications. 2015; **51**(88):15894-15897

[50] Narayanaswamy K, Swetha T, Kapil G, Pandey SS, Hayase S, Singh SP. Simple metal-free dyes derived from triphenylamine for DSSC: A comparative study of two different anchoring group. Electrochimica Acta. 2015;**169**:256-263

[51] Massin J, Ducasse L, Abbas M, Hirsch L, Toupance T, Olivier C. Molecular engineering of carbazole-fluorene sensitizers for high open-circuit voltage DSSCs: Synthesis and performance comparison with iodine and cobalt electrolytes. Dyes and Pigments. 2015;**118**:76-87

[52] Capodilupo AL, De Marco L, Corrente GA, Giannuzzi R, Fabiano E, Cardone A, et al. Synthesis and characterization of a new series of dibenzofulvene based organic dyes for DSSCs. Dyes and Pigments. 2016;**130**: 79-89

[53] Demeter D, Melchiorre F, Biagini P, Jungsuttiwong S, Po R, Roncali J. 3, 4-Ethylenedioxythiophene (EDOT) and 3, 4-ethylenedithiathiophene (EDTT) as terminal blocks for oligothiophene dyes for DSSCs. Tetrahedron Letters. 2016; **57**(43):4815-4820

[54] Liao Y, Hu J, Zhu C, Liu Y, Xu C, Chen C, et al. Synthesis and photovoltaic performance of novel polymeric metal complex sensitizer with benzodithiophene or carbazole derivative as donor in dye-sensitized solar cell. Journal of Molecular Structure. 2016;**1108**:467-474

[55] Park K-W, Serrano LA, Ahn S, Baek MH, Wiles AA, Cooke G, et al. An investigation of the role the donor moiety plays in modulating the efficiency of 'donor-π-acceptor-π-acceptor'organic DSSCs. Tetrahedron. 2017;**73**(8):1098-1104

[56] Chiu KY, Tran TTH, Chang SH, Yang T-F, Yuhlong Oliver S. A new series of azobenzene-bridged metal-free organic dyes and application on DSSC. Dyes and Pigments. 2017;**146**: 512-519

[57] Elmorsy MR, Rui S, Fadda AA, Etman HA, Tawfik EH, El-Shafei A. Molecular engineering and synthesis of novel metal-free organic sensitizers with D-π-A-π-a architecture for DSSC applications: The effect of the anchoring group. Dyes and Pigments. 2018;**158**: 121-130

[58] Pati PB, Yang W, Zade SS. New dyes for DSSC containing triphenylamine based extended donor: Synthesis, photophysical properties and device performance. Spectrochimica Acta Part A: Molecular and Biomolecular Spectroscopy. 2017;**178**:106-113

[59] Fattori A, Majer R, Mazzanti A, Ottaviani MF, Modelli A, Mantellini F, et al. N-heterocyclic linkers from 1, 2-diaza-1, 3-dienes for dye-sensitized solar cells: DFT calculations, synthesis and photovoltaic performance. Dyes and Pigments. 2017;**145**:246-255

[60] Zhang K, Zhang W, Huang J, Pang A, Wong MS. Metal-free photosensitizers based on benzodithienothiophene as π-conjugated spacer for dye-sensitized solar cells. Organic Electronics. 2017;**42**:275-283

[61] Rajavelu K, Sudip M, Kothandaraman R, Rajakumar P. Synthesis and DSSC application of triazole bridged dendrimers with benzoheterazole surface groups. Solar Energy. 2018;**166**:379-389

[62] Fernandes SSM, Pereira A, Ivanou D, Mendes A, Manuela M, Raposo M. Benzothiadiazole derivatives functionalized with two different (hetero) aromatic donor groups: Synthesis and evaluation as TiO_2 sensitizers for DSSCs. Dyes and Pigments. 2018;**151**:89-94

[63] Naik P, Babu DD, Rui S, El-Shafei A, Adhikari AV. Synthesis, characterization and performance studies of a new metal-free organic sensitizer for DSSC application. Materials Today: Proceedings. 2018; **5**(1):3150-3157

[64] Suarez M, Carolina C, Jimmy M, Edwin F-L, Yenny Á-T. Design, theoretical study and correlation of the electronic and optical properties of diethynylphenylthiophene as photovoltaic materials. Journal of Molecular Structure. 2020;**1201**:127093

Chapter 9

Coordination States and Catalytic Performance of Ti in Titanium Silicalite-1

Yi Zuo, Min Liu and Xinwen Guo

Abstract

In the past two decades, we studied the synthesis, modification, and application of titanium silicalite-1 (TS-1) systematically with the goal of exploring its role as a catalyst for industrial selective oxidation reactions. Three factors were primarily considered for catalytic performance: the coordination states of titanium ions, locations of titanium ions, and diffusion properties. The coordination state of Ti, which was the most important of all the three factors, was tuned by controlling the synthesis conditions and posttreating with organic bases. Spectroscopy was used to help establishing the relationship between catalytic activity and coordination state. More active titanium species were located on the external surface by posttreatment, so the catalytic performance for larger molecules was improved significantly. The diffusion properties can be enhanced by posttreatment with organic bases. Furthermore, meso−/microporous titanium silicalite was synthesized by one-pot synthesis with cetyltrimethyl ammonium bromide (CTAB) as a mesoporous porogen. The TS-1 plate with a short *b*-axis length was also provided.

Keywords: coordination state, titanium, stability, titanium silicalite-1, selective oxidation

1. Introduction

Natural zeolites are a kind of hydrated crystalline silica-aluminate with specific pore structure. The pore diameters of zeolites are similar to the sizes of molecules. Because they can sieve molecules, they are also known as molecular sieves. According to the pore diameter, IUPAC classified porous materials into microporous materials (<2 nm), mesoporous materials (2–50 nm), and macroporous materials (>50 nm) [1]. Zeolites are a microporous material. Due to their high hydrothermal stability, simple synthesis process, strong adsorption properties, adjustable acidity and alkalinity, and pore shape selectivity, they are widely used in petroleum refining, the chemical industry, and separations [2]. The most commonly used zeolites, such as type-A, faujasite, mordenite, and ZSM-5, have narrow pore size and more serious diffusion restrictions, which are disadvantageous. Mesoporous materials have larger surface areas and pore volumes, which is more favorable for larger molecular reactions, adsorption, and separation. However, the hydrothermal stability is not satisfactory due to the amorphous silica wall of mesopores.

Titanium silicalite-1 (TS-1) with MFI topology was first hydrothermally synthesized by Taramasso et al. [3]. After that, it has attracted much attention due to its excellent catalytic activity for selective oxidation with H_2O_2, such as alkene epoxidation [4–9], aromatics hydroxylation [10–12], ketone ammoximation [13–15], alkane oxidation [16, 17], and so on [18, 19]. Therefore, it is considered a milestone in the field of zeolitic catalysis. MFI topology contains two types of 10-membered ring channels, which are the straight channel (0.56 × 0.54 nm) and zigzag channel (0.55 × 0.51 nm). The substitution of titanium atoms for framework silicon or aluminum atoms generates a molecular sieve with tetrahedrally coordinated Ti. The isolated tetrahedrally coordinated titanium (also called framework Ti) in TS-1 is the main active center for catalytic oxidation. However, the amount of tetrahedrally coordinated Ti is limited (2.5 mol%), because the lattice expansion inhibits the insertion of Ti into the framework [20]. In the next 30 years, phenol hydroxylation to benzene diols, cyclohexanone ammoximation to cyclohexanone oxime, butanone ammoximation to diacetyl monoxime, and propene epoxidation to propene oxide catalyzed by TS-1 were commercialized successively. Nevertheless, there are still many problems in the synthesis and application of TS-1, such as the transformation of tetrahedrally coordinated Ti to octahedrally coordinated Ti or anatase TiO_2 (loss of the active center) in the reaction, and the deactivation of the catalyst by blocking of channels [21]. Therefore, many researchers have made an effort to solve these problems, and so do we. To further improve the catalytic performance and expand the application of TS-1, it is necessary to summarize our current research achievement.

In this chapter, we describe our recent progress on controlling Ti coordination states, design of porosity, and applications of TS-1. We hope that this summary will help in understanding the developing process and our contribution to research on TS-1.

2. Controlment of the Ti coordination states

2.1 Tuning of synthesis conditions

There are mainly three kinds of Ti species in TS-1. Tetrahedrally coordinated titanium, which was mentioned above, is usually considered to be the active center for the oxidation reactions. This Ti species can form a five-membered ring (5MR) structure with hydrogen peroxide and alcohol (see **Figure 1**) [22], which can oxidize propene to prepare propene oxide. The 5MR structure is very stable and is easy to be formed in the epoxidation of alkenes, according to density functional theory study. The β-O atom in the structure is partially positive due to the hydrogen bond between alcohol and O-O-H; thus, the C=C bond can react with the β-O atom by nucleophilic attack. The 5MR mechanism explains the effects of alcohol solvents

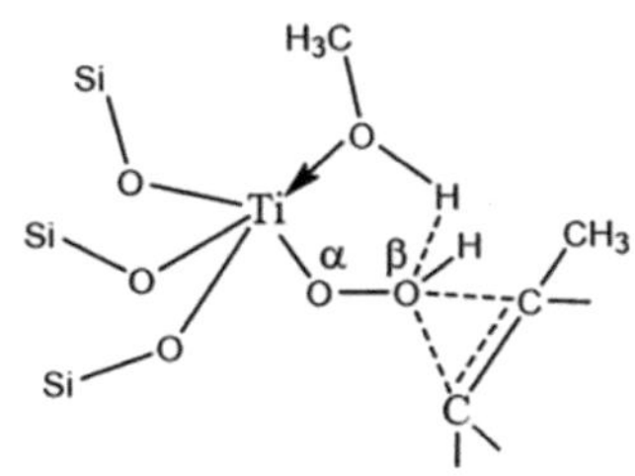

Figure 1.
Diagram of 5MR structure.

in propene epoxidation well; hence it is widely accepted. The other two types of Ti species in TS-1 are octahedrally coordinated Ti (usually called extra-framework Ti) and anatase TiO_2. When the feeding amount of titanium source in the synthesis gel is more than 2.5 mol%, the excessive Ti will transform to these two Ti species. The anatase TiO_2 can decompose hydrogen peroxide to water and O_2; thus, the generation of anatase TiO_2 should be avoided [23]. The octahedrally coordinated Ti can be generated from the tetrahedrally coordinated Ti coordinating with two water molecules. The function of octahedrally coordinated Ti is still controversial. Former researchers thought it was inert for oxidation, but recent works have reported that it was also active for selective oxidation [24]. Wang et al. found that octahedrally coordinated Ti is an intermediate in propene epoxidation [25].

It is well known that the hydrolysis rate of titanium sources is faster than that of silicon sources and the similar hydrolysis extent or crystallization rates of them benefit for the generation of more tetrahedrally coordinated Ti. Therefore, we hydrolyzed the two sources individually to complete hydrolysis simultaneously [26]. Tetrapropylammonium hydroxide (TPAOH) was used as template and base and was added to silicon and titanium sources. The silicon source was hydrolyzed at 313 K for 5 h, while the titanium source was at room temperature for 0.5 h. The hydrolysis of the two sources was completed at the same time. After that, the two hydrolysates were mixed together and crystallized at 443 K for 2 d. Under these conditions, the content of tetrahedrally coordinated Ti in TS-1 was ~1 wt%, while the total content of Ti was ~1.9 wt%. From this result, we know that the generation of tetrahedrally coordinated Ti is difficult.

Until now, tetrapropylammonium ions (TPA^+) are necessary for synthesizing TS-1 as the template. In consideration of the high price of TPAOH, many researchers tend to use tetrapropylammonium bromide (TPABr) to structure-directly synthesize TS-1 because they have the same TPA^+ cation, but TPABr is much cheaper. However, due to the introduction of Br^- and the reduction of basicity, the particle size of TS-1 enlarges obviously when TPABr is used. The TPAOH system often obtains nanosized TS-1, while the TPABr system usually gets microsized particles. We explored a method for synthesizing small-crystal TS-1 in the TPABr system [27], which will be presented in detail in Section 3.1. Herein, we only discuss the influence of molar ratio of Si/Ti (n(Si/Ti)) on the coordination states of Ti when synthesizing small-crystal TS-1 in the TPABr system. Small-crystal TS-1 with different feeding n(Si/Ti) (20, 50, and 80) was synthesized by adding different amounts of titanium source ($TiCl_4$) to the synthesis gel. The weight content of silicon and titanium in the samples obtained by inductively coupled plasma-optical emission spectrometer (ICP-OES) shows that the actual n(Si/Ti) was slightly higher than the feeding one, except for sample with the feeding n(Si/Ti) of 20. Ultraviolet/visible diffuse reflectance (UV/vis) spectroscopy, Raman spectroscopy, and X-ray adsorption near edge structure (XANES) spectroscopy were used to study the coordination states of Ti qualitatively and quantitatively [28].

UV/vis spectroscopy is one of the first spectral techniques used for the detection of Ti coordination states in titanium silicalites. Peak deconvolutions were performed using the PeakFit program with the Gaussian fitting method. In the spectra of small-crystal TS-1 (**Figure 2**), there are three major absorption bands centered at 200–210, 230–290, and 310–330 nm. The band at 200–210 nm is assigned to tetrahedrally coordinated Ti, while that at 310–330 nm belongs to anatase TiO_2. There are more than one kind of Ti species between 230 and 290 nm in the UV/vis spectra of TS-1. The band at approximately 250–290 nm is attributed to the octahedrally coordinated Ti species, which is inactive for the oxidation reactions, and the band at 230–250 nm is an isolated Ti species with a lower coordination number of oxygen than octahedrally coordinated Ti (such as pentahedrally coordinated Ti). The undercoordinated Ti has a higher energy than the octahedral one, so its band shifts to a shorter wavelength. The content of tetrahedrally coordinated Ti, octahedrally coordinated Ti, and anatase TiO_2 increases when *n*(Si/Ti) is decreased, and the increase of anatase TiO_2 content is much stronger than that of other species, proving that the introduction of Ti in the framework is limited. It is notable that a new band appears at ~235 nm in the sample with *n*(Si/Ti) of 80. We consider it to belong to the pentahedrally coordinated Ti by combining these results with the results of XANES. This Ti state might be formed by one Ti atom with five "SiO_4". Thus, a high *n*(Si/Ti) promotes more "SiO_4" coordination with this Ti atom, which is beneficial to the generation of pentahedrally coordinated Ti.

The catalytic performance of the small-crystal TS-1 with different *n*(Si/Ti) was evaluated in the epoxidation of propene. The conversion of H_2O_2 over TS-1 with *n*(Si/Ti) of 20 is the lowest, indicating that the Ti coordination state is more important than its content for the catalytic activity. The highest turnover frequency (TOF) was obtained from the TS-1 with *n*(Si/Ti) of 80, demonstrating that the pentahedrally coordinated Ti was the most active species of all the Ti coordination types. The lowest TOF was obtained over the TS-1 with *n*(Si/Ti) of 20, indicating that octahedrally coordinated Ti was inert or had negative effects on the epoxidation. Based on the XANES and TOF data, we calculated the contents of differently coordinated Ti in the three samples (see **Table 1**). The content of pentahedrally coordinated Ti decreases and the content of octahedrally coordinated Ti increases with the decrease of feeding *n*(Si/Ti) from 80 to 50, suggesting that the insertion of Ti into the framework is at nearly the maximum at the *n*(Si/Ti) of 80. Continuing to add Ti leads to a sharp increase of octahedrally coordinated Ti. The TOFs of pure tetrahedrally, pentahedrally, and octahedrally coordinated Ti in small-crystal TS-1, which were calculated according to the total TOFs and contents of differently coordinated Ti, are 373.3, 1434.3, and 0 mol H_2O_2/(h·mol Ti), respectively. These results confirm that pentahedrally coordinated Ti is the most active species among the three coordination structures.

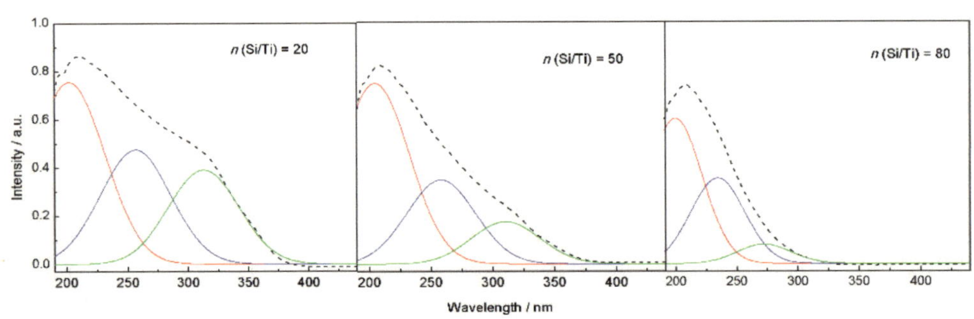

Figure 2.
UV/Vis spectra of the TS-1 with different n*(Si/Ti).*

n(Si/Ti)	Percentage of differently coordinated Ti/%		
	Tetrahedral Ti	Pentahedral Ti	Octahedral Ti
20	42	0	58
50	76	21	3
80	50	50	0

Table 1.
Contents of differently coordinated Ti.

Nevertheless, the controllable synthesis of TS-1 with a large amount of pentahedrally coordinated Ti is still impossible. When more Ti was added to the synthesis gel, the pentahedrally coordinated Ti would transform to tetrahedrally and octahedrally coordinated Ti due to the loss of pentahedrally coordinated Ti generation conditions mentioned previously.

In addition, the stability sequence of differently coordinated Ti is anatase TiO_2 > octahedrally coordinated Ti > tetrahedrally coordinated Ti > pentahedrally coordinated Ti. The catalytic activity sequence is opposite to the stability, because unstabilization means high energy and thus high catalytic activity.

2.2 Usage of additives

From the above introduction, we know that matching the crystallization rates of silicon and titanium sources benefits the generation of tetrahedrally coordinated Ti. Therefore, some researchers tried to control the crystallization process by adding some modifiers. Fan et al. used different ammonium salts as the crystallization-mediating agents to synthesize TS-1 [29]. They found that the ammonium salts could not only drastically decrease the pH of the synthesis gel and slow down crystallization, but they could also modify the crystallization mechanism and make the incorporation of titanium into the framework match well with that of silicon. As a result, the formation of octahedrally coordinated Ti and anatase TiO_2 was eliminated successfully. It was reported that the anionic polyelectrolyte poly(acrylic acid) was also able to facilitate the insertion of Ti to the framework via a liquid-phase and solid-phase transformation mechanism [30]. Some researchers used sucrose as the modifier [31], which would carbonize during the crystallization of TS-1 and release hydrogen ions. Therefore, the pH of the hydrothermal system was reduced, and the sucrose played a similar role to the ammonium salts.

We adopted two natural macromolecular additives to adjust the coordination states of Ti, which were starch and gelatin [32]. When they were introduced to the synthesis gel, they also released hydrogen ions as the sucrose did. Hence, they can also tune the coordination states of Ti.

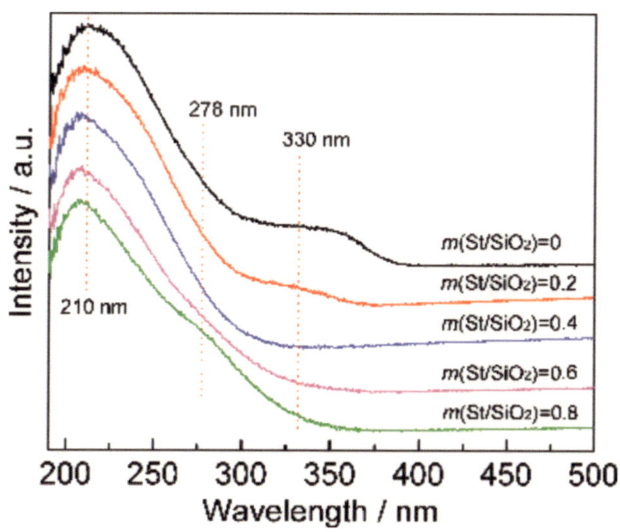

Figure 3.
UV/Vis spectra of the TS-1 synthesized with different amounts of starch.

The addition of starch hardly influences the morphology of TS-1 but can eliminate the formation of extra-framework Ti. **Figure 3** shows the UV/vis spectra of TS-1 synthesized with different amounts of starch. An obvious absorption band at ~330 nm appears in the TS-1 synthesized without starch, proving the existence of anatase TiO_2. As the content of starch and gelatin increases, the band of anatase TiO_2 disappears gradually. The content of octahedrally coordinated Ti decreases with the increase of starch until the weight ratio of starch/SiO_2 (m(St/SiO_2)) reaches 0.6. Then, the content of octahedrally coordinated Ti increases slightly, probably due to the introduction of starch promoting the coordination saturation of titanium ions. Compared to the TS-1 synthesized without starch, those obtained with starch have a higher content of tetrahedrally coordinated Ti, a quite low content of octahedrally coordinated Ti, and are free of anatase TiO_2. Therefore, they show a much higher catalytic activity for the epoxidation of 1-butene.

Gelatin is similar to starch in its effect on Ti coordination states. However, it contains both amino and carboxyl groups. Thus, it has the ability to tune the morphology of MFI-typed zeolites. This will be further discussed in Section 3.4.

2.3 Posttreatment with organic bases

Many studies focused on the treatment of zeolites with organic bases, especially for the quaternary ammonium bases, because the treatment could improve the catalytic performance significantly. The treatment leads to the dissolution of "SiO_4" in the TS-1 crystals and recrystallization on the external surface of crystals, generating hollow zeolites, which decreases the diffusion resistance. When the "SiO_4" was dissolved, the coordination states of the neighbored Ti ions would be changed accordingly. Two Si-O bonds near to the tetrahedrally coordinated Ti may be broken, and the tetrahedrally coordinated Ti may transform to octahedrally coordinated Ti after combining with two water molecules (**Figure 4**).

We studied the treatment of small-crystal TS-1 with different organic bases, including ethylamine (EA), diethylamine (DEA), tetramethylammonium hydroxide (TMAOH), and tetrapropylammonium hydroxide (TPAOH) solutions [33]. The catalytic performances of phenol hydroxylation over the treated samples were improved to different extents. The TS-1 treated with TPAOH has the highest catalytic activity in the treated samples.

To understand the reason for this result, we characterized the treated samples with Ti L-edge XANES spectroscopy, the spectra of which are shown in **Figure 5**. The spectra consist of two sets of doublets, which correspond to the $2p_{1/2}$ and $2p_{3/2}$ transitions of the $3d^0$ to $2p^5 3d^1$ states. The $L2$ edge is at a higher energy (462–470 eV), and the $L3$ edge is at a lower energy (455–462 eV). The splitting of each edge is attributed to the t_{eg} and e_g symmetry of the d orbital.

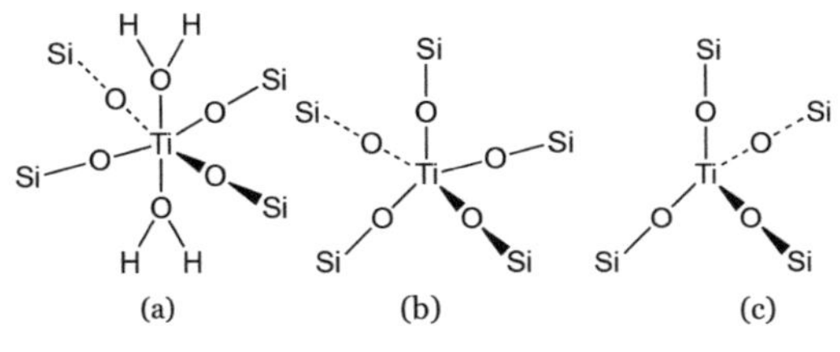

Figure 4.
Structures of octahedrally coordinated Ti (a), pentahedrally coordinated Ti (b), and tetrahedrally coordinated Ti (c).

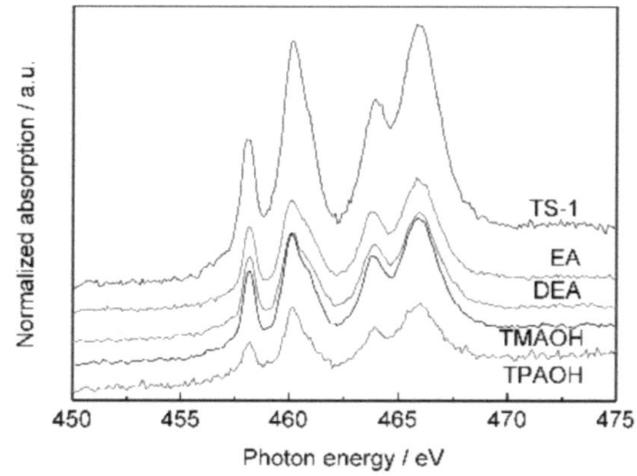

Figure 5.
Ti L-edge XANES spectra of the TS-1 treated with different bases.

The higher energy peak of the $L3$ edge (459–462 eV) consists of two peaks in some samples, a main peak and a shoulder one, which is on the lower energy side (~460 eV) for rutile TiO_2 and the higher energy side (461 eV) for anatase TiO_2. The intensities of the two peaks are reversed for the substances that they represent. The tetrahedrally coordinated Ti is characterized by the absence of splitting of the peak at 459–462 eV and a relatively weaker intensity of the peaks at lower energies of both the $L2$ and $L3$ edges than those at higher energies. The pentahedrally coordinated Ti is characterized by a slight shift to higher energy and a drastic decrease of the lower energy peak of $L3$, a shift of the higher energy peak of $L2$ to lower energy, and the appearance of a shoulder peak on the lower energy peak of $L2$. We found that pentahedrally coordinated Ti existed in the TPAOH-treated TS-1, but it was absent in the other samples. Therefore, the generation of pentahedrally coordinated Ti is another reason for the increasing activity of TPAOH-treated TS-1. The possible structure of pentahedrally coordinated Ti is shown in **Figure 4**, the stable form of which is tetragonal pyramid.

3. Improvement of the diffusion property

3.1 Reducing of particle size in TPABr system

Most reactions catalyzed by zeolites occur in their channels. A short channel means a short diffusion pathway for reactants from bulk to active centers (such as tetrahedrally coordinated Ti), therefore reducing the particle size benefits the diffusion. We have mentioned in Section 2.1 that the particle size of TS-1 obtained in the TPABr hydrothermal system is often at the micron scale, which is disadvantageous for diffusion. Hence, we tried to control the particle size of TS-1 in the TPABr system by adding different seeds. First, we used the mother liquor of nanosized TS-1 as the seed [27]. The synthesis process is illustrated in **Figure 6**. The mother liquor was prepared by crystallizing the synthesis gel at 443 K for 48 h, according to prior work [26]. The size of the obtained seed is ~100 nm. When using powdery TS-1 as the seed, microsized TS-1 was obtained, the size of which was 2 × 1 × 0.5 μm. However, when the seed was changed to the mother liquor, the size decreased significantly to 600 × 400 × 250 nm, so we called it small-crystal TS-1. Its catalytic performance was evaluated in the epoxidation of propene and hydroxylation of phenol. The conversion of H_2O_2 and selectivity of

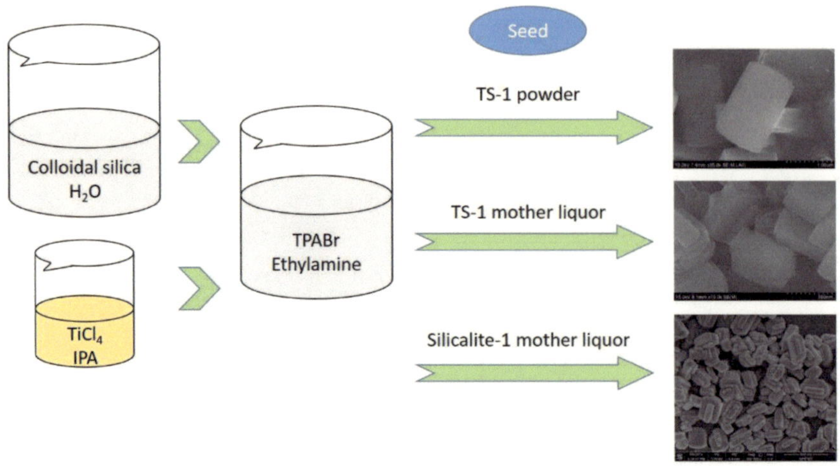

Figure 6.
Synthesis process of TS-1 in TPABr system.

PO on small-crystal TS-1 reached 92 and 98%, respectively. In the hydroxylation of phenol, using small-crystal TS-1 as a catalyst also resulted in a higher conversion of phenol.

The synthesis conditions were then systematically studied, including purification methods for small-crystal TS-1, the Si/Ti molar ratio in TS-1, the amount of the seed, and the crystallization period [34]. It was found that the size of TS-1 was not significantly affected by the synthesis conditions except for the amount of seed. The highest catalytic performance of the propene epoxidation was obtained when the TS-1 was purified three times by precipitation; the n(Si/Ti) was 50, the weight ratio of seed/SiO_2 was 0.06, and the crystallization time was 48 h.

After that, the seed was changed to the mother liquor of nanosized silicalite-1, the size of which was ~80 nm [35]. The particle size can be adjusted from 1200 to 200 nm by varying the seed amount from 0.05 to 12 wt%. The relationship between particle size and the content of seed is shown in **Figure 7**. The data fits well to a power function, with the fit equation of:

$$\text{Crystal size} = 468.7\ (\text{content of seed})^{-0.3158} \tag{1}$$

The degree of fitting measured by the coefficient of determination (R^2) is 0.9968. Crystallization time plays a less important role than seed amount on the adjustment of the size. TS-1 with different crystal sizes was characterized and evaluated in the propene epoxidation. The catalytic activity and selectivity for PO are enhanced by decreasing the particle size from 1200 to 200 nm, due to gradually eliminated diffusion limitations. The seed is significant for this system, because eliminating the seed leads to poor crystallization and catalytic activity. The mechanism of the seed function was studied by simulating the transformation process of the seed in the TS-1 synthesis system. When the content of seed is lower than 1 wt%, it primarily performed as a nucleus for the growth of silicon and titanium sources. As the content increases, more seed will dissolute to secondary structural units first and then accelerate the crystallization.

Furthermore, the low pH of the TPABr system promotes the similarity of crystallization rates of silicon and titanium sources, inhibiting the generation of

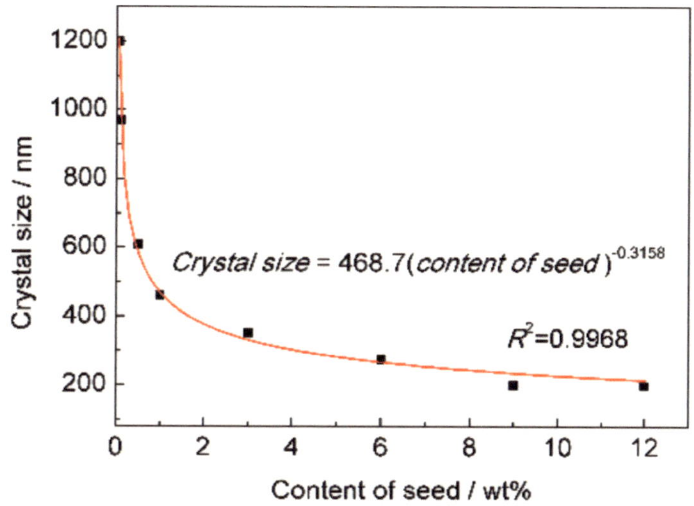

Figure 7.
Particle size of TS-1 plotted against the weight percent of seed using in synthesis. (data points represent experimental data, and the continuous curve represents a power law fit to the data with equation given on the plot).

octahedrally coordinated Ti and anatase TiO_2. The content of tetrahedrally coordinated Ti achieved ~1.2 wt% when the feeding n(Si/Ti) was 50.

3.2 Design of hollow TS-1 materials

In addition to reducing the particle size, synthesis of hollow materials can also enhance diffusion properties. Treating with TPAOH solution is one of the most commonly used methods for generating hollow spaces in TS-1. However, this method leads to the transformation of tetrahedrally coordinated Ti to extra-framework Ti, which is harmful for the catalytic performance. Therefore, we provided a method for synthesizing a hollow core-shell material to prevent the generation of extra-framework Ti [36]. The hollow silicalite-1@titanium silicalite-1 (H-S-1@TS-1) core-shell material was synthesized in a TPAOH hydrothermal system with hollow silicalite-1 serving as the core (see **Figure 8**). The very small TS-1 particles grow along the external surface of hollow silicalite-1, thus generating hollow material. Since the hollow structure was given by silicalite-1, the tetrahedrally coordinated Ti in H-S-1@TS-1 was not converted to other coordination states. Thus, the extra-framework Ti was absent in H-S-1@TS-1. Due to the synergy function of pure tetrahedrally coordinated Ti species, higher Ti content on external surface, and enhanced diffusion properties, H-S-1@TS-1 showed better propene epoxidation activity (TOF = 13.61 mol/(mol·h)) than traditional microporous TS-1 (TOF = 8.08 mol/(mol·h)) and that posttreated with TPAOH solution (TOF = 9.75 mol/(mol·h)).

3.3 Insertion of Ti on the external surface

In the posttreatment with TPAOH solution, the "SiO4" in the crystals is dissolved and recrystallized on the external surface. When TS-1 was extruded with silica as the support and the obtained extrudate was treated with TPAOH solution, the silica support would crystallize and restrain the dissolution of "SiO_4" in the crystals [37]. Hence, we introduced a titanium source to the posttreatment to make it crystallize

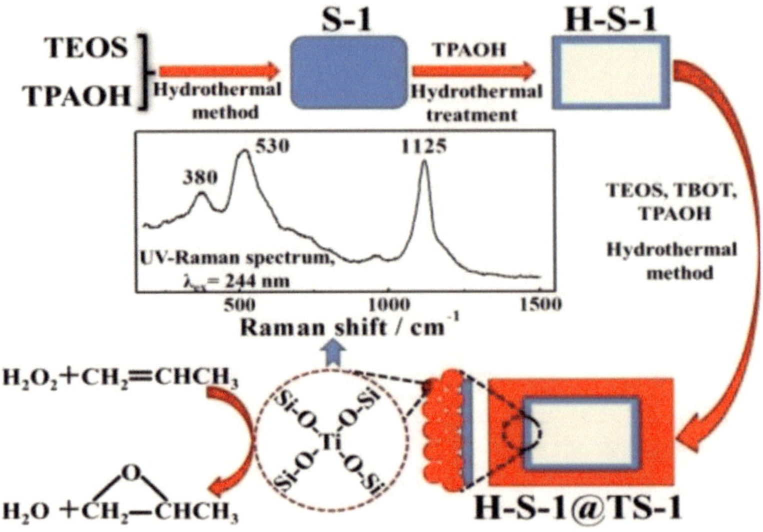

Figure 8.
Preparation process of H-S-1@TS-1.

Ti content/vol%	In the bulk			On the surface		
	SiO_2/wt%	TiO_2/wt%	*n*(Si/Ti)	SiO_2/wt%	TiO_2/wt%	*n*(Si/Ti)
Untreated	97.06	2.79	46.4	98.67	1.18	111.9
0	97.10	2.76	46.9	98.99	0.86	153.5
5	96.96	2.89	44.8	98.51	1.34	98.1
10	96.81	3.02	42.7	98.29	1.54	85.3
16	96.73	3.12	41.4	98.28	1.57	83.6
25	96.65	3.20	40.3	98.25	1.60	81.7

[a]*The elemental compositions in the bulk and on the surface were determined by ICP-OES and XPS, respectively.*

Table 2.
Elemental composition of the TS-1 treated with different Ti contents[a].

with silica support and insert more Ti on the external surface [38]. Different amounts of the titanium source (tetrabutyl titanate hydrolysate) were added to the TPAOH solution. **Table 2** shows the elemental compositions of the samples both in the bulk and on the external surface. The *n*(Si/Ti) in the bulk is hardly affected by the individual TPAOH treatment. The Ti content increases gradually with increasing concentration of Ti in the postsynthesis solution. However, not all Ti in the samples transforms to tetrahedrally coordinated Ti. The X-ray photoelectron spectroscopy (XPS) results show that the content of Ti on the external surface reduced after the treatment with TPAOH solution, which is probably due to the crystallization of amorphous silica occurring at the external surface of TS-1 particles and covering more Ti species. The *n*(Si/Ti) on the external surface decreases as Ti concentration in the postsynthesis solution increases, indicating that Ti can be located at the external surface under these postsynthesis conditions, and the amount of Ti is restricted by the Ti concentration and the limitation of Ti in the MFI topology. The samples were evaluated in the hydroxylation of phenol. The conversion of phenol increases to a different extent after the postsynthesis. The hollow spaces generated in the crystals reducing the diffusion limitation may be one of the reasons for the improvement of catalytic activity. The new generated tetrahedrally coordinated Ti on the external surface also provides more easily adsorbable active centers for reactants. At low Ti concentrations (≤10 vol%), increasing the Ti amount generates more active Ti centers on the external surface. Thus, the conversion of phenol increases as Ti concentration increases. However, at high Ti concentrations, the excessive Ti is converted to anatase TiO_2 due to the limitation of titanium content in the MFI topology. The anatase TiO_2 can block the channels and cover the active centers. Thus, the conversion decreases as Ti concentration increases over 10 vol%.

3.4 Preparation of plate-like TS-1

We have mentioned in Section 2.2 that adding gelatin to the synthesis gel of TS-1 could modify its morphology. Actually, it can adjust the thickness of the *b*-axis of MFI-type zeolites. The *b*-axis direction is parallel to the straight channel, so the short, straight, and open channels of zeolites with MFI topology are tailored for diffusion and catalysis. Ti-MFI, Al-MFI, Zr-MFI, Mn-MFI, Cu-MFI, and Fe-MFI plates were synthesized with *b*-axis lengths ranging from 40 to 200 nm by these means. The lengths of the other two dimensions are submicron-sized, which leads to an easy separation of zeolites from the mother liquor. The synergistic effects of the amino and carboxyl groups in gelatin lead to the generation of plate-like zeolites. The physical adsorption of cyclohexane indicates that TS-1 with a thickness of 40 nm has a faster diffusion rate than that of the traditional aggregated material. The TOF of cyclohexene epoxidation over TS-1 plates is about four times that of traditional nanosized TS-1.

3.5 One-pot synthesis of meso−/microporous titanium silicalite

In recent years, research on the synthesis of hierarchical molecular sieves has attracted much attention, because they have the advantages of microporous (good catalytic activity and hydrothermal stability) and mesoporous (excellent diffusion property) materials simultaneously. The preparation methods of hierarchical molecular sieves are mainly postsynthesis and one-pot synthesis. The postsynthesis method uses acid or base to treat micropores, which was introduced above. One-pot synthesis of hierarchical titanium silicalite was first reported by Jacobsen et al. using carbon black as a hard template [39]. After that, a series of hierarchical titanium silicalites were synthesized by using different carbon-based materials. However, the complexity of the synthesis procedure seriously limited the industrial applications of these hierarchical materials. An attractive method is to utilize suitable surfactants as soft mesoporous templates for the direct synthesis of hierarchical materials. Cetyltrimethyl ammonium bromide (CTAB) is one of the most commonly used surfactants. Nevertheless, the mesopores in the CTAB-directed materials are mostly intercrystals. Furthermore, the micropores and mesopores are often phase-separated from each other. We explored an easy and new route for synthesizing the meso−/microporous titanium silicalite with controllable pore diameter by using CTAB and TPAOH as mesoporous and microporous template, respectively [40]. The new route is adding CTAB to the hydrolysis reaction of the silicon source so it forms mesopores prior to the crystallization of microporous MFI topology and prevents the occurrence of phase separation. In other words, this porosity formation sequence makes the two kinds of channels in the materials, which are micropores with MFI topology and mesopores with worm-like morphology, distributed homogeneously. The pore diameter of the mesopores can be adjusted from the maximum

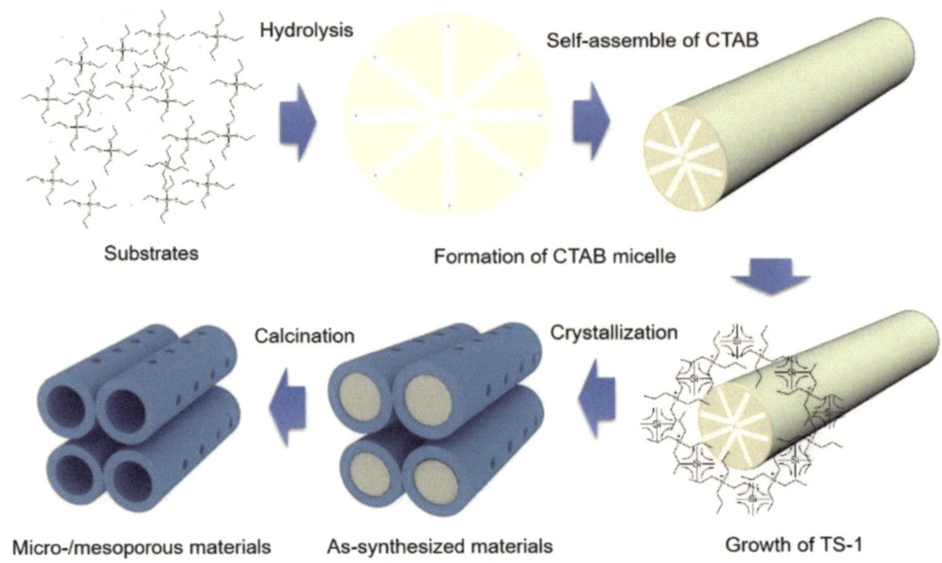

Figure 9.
Synthesis process of meso−/microporous titanium silicalite.

center of 2.6 nm to that of 6.9 nm by tuning the molar ratio of CTAB to silicon from 0.125 to 0.20. The introduction of CTAB also causes the variation in coordination states and location of Ti ions in the materials. More CTAB leads to a higher content of octahedrally coordinated Ti and a lower content of tetrahedrally coordinated Ti. Furthermore, more Ti is located near the external surface of TS-1 crystals, when adding more CTAB to the synthesis gel.

The meso−/microporous titanium silicalite catalysts were evaluated in the epoxidation of cyclohexene and showed excellent catalytic activity with respect to the conventional microporous TS-1, due to the enhanced diffusion properties in the mesopores and higher titanium content near the external surface of the former (**Figure 9**).

4. Industrial application of HPPO route

PO is an important organic chemical intermediate among propene derivatives. Most PO is used to produce polyether polyols and polyurethane. Since 2003, the consumption of PO in the world has been increasing year by year. There are about 20 routes for PO production, among which the Chlorohydrin and Halcon routes are the most commonly used. The investment cost of Chlorohydrin route is low, but it produces a large amount of wastewater containing Cl^-, which pollutes the environment and seriously corrodes equipment. The Halcon route overcomes the disadvantages of environmental pollution, but the cost is high. Moreover, the economy of PO is seriously affected by the coproducts. Therefore, PO manufacture needs a new route.

At present, hydrogen peroxide to propene oxide (HPPO) route is one of the most possible alternatives for PO production. Compared to the traditional routes, the HPPO route provides environmental and economic benefits. In recent years, the HPPO route was commercialized by BASF/Dow Chemical and Evonik/Uhde in Belgium and South Korea, separately. Some institutes also tried this route in pilot plants. The main and side reactions in HPPO route are shown in **Figure 10**. It is clear that all the reactions are exothermal. The exothermic reaction not only

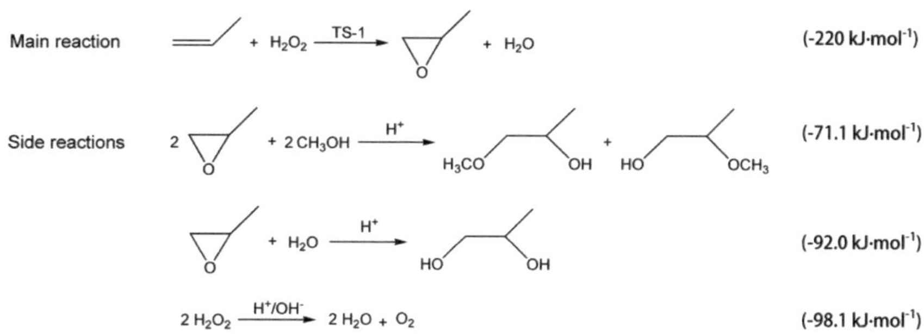

Figure 10.
Main and side reactions in HPPO route.

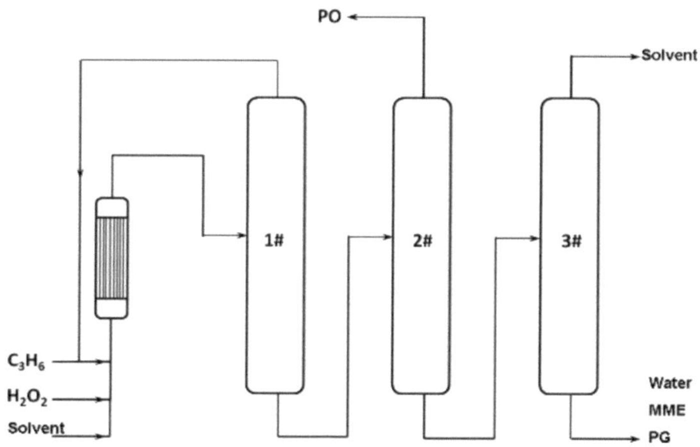

Figure 11.
Flow chart of the 100 t/a HPPO pilot plant.

threatens safety but also promotes the solvolysis reactions and generates PO oligomers. The blocking of TS-1 channels by oligomers is the main cause for its deactivation.

Three steps were taken for the industrialization of our HPPO route, which were the 100 t/a pilot plant, 1000 t/a pilot plant, and 150 kt/a industrial plant. In 2009, the 100 t/a pilot plant procedure was carried out in Jiangsu, China, the flow chart of which is shown in **Figure 11**. A fixed-bed reactor was adopted, and the loading of the catalyst was 100 kg. Propene, H_2O_2, and solvent were fed into the reactor simultaneously by three pumps. The product flowed out of the reactor and entered the 1# rectifying column, in which propene was separated from the top of the column. Then, propene flowed to the propene storage tank for recycling. The materials from the bottom of the 1# rectifying column entered the 2# rectifying column. PO was separated from the top of this column and entered the finished tank. The materials from the bottom of the 2# rectifying column entered the 3# rectifying column. The solvent was separated from the top of the column and put into the solvent storage tank for recycling. The material in the bottom of the column contained mainly water and a small amount of MME and PG.

The 1000 t/a pilot plant and 150 kt/a industrial plants used similar technology to the 100 t/a pilot plant, except for some energy optimization. The former was carried out in 2013, while the latter is under construction. Under the optimized reaction conditions, the conversion of H_2O_2 and selectivity of PO are both higher than 95%, and the purity of PO is more than 99.95% in the three HPPO routes.

The industrial catalyst deactivated partly after the 100 t/a pilot plant reaction. The activity decreased from the inlet to the outlet of the pilot plant reactor [41]. The main reason for the deactivation in pilot plant is similar to that in laboratory, which is blocking of pores and covering of active centers by ethers or oligomers. The more oligomers are generated, the more seriously the catalyst deactivates. The loss of a small amount of framework titanium had little influence on the catalytic activity.

The deactivated catalysts could be externally regenerated by calcination at 813 K for 6 h and in situ regenerated by washing with dilute H_2O_2. The in situ regeneration, rather than the external regeneration, can be adopted in industry. When using in situ regeneration with dilute H_2O_2, a longer washing time is more effective than a higher concentration of H_2O_2. There is one thing needing to be concerned for in situ regeneration. If the concentration of H_2O_2 is too high or the washing time is too long, some tetrahedrally coordinated Ti will be leached out and transform to octahedrally coordinated Ti.

5. Conclusions

Great progress has been made in the synthesis of TS-1 and improvement of its catalytic properties. The relative technologies have become increasingly more mature. However, the active center in TS-1 is still controversial (tetrahedrally, pentahedrally, and/or octahedrally coordinated Ti). The contradiction between the cost and catalytic performance of TS-1 has not been resolved completely. Therefore, the synthesis of high-performing, low cost TS-1 needs to be further studied. At present, people attach great importance to environmental protection; thus, the application of TS-1 will have broad development.

This chapter summarized recent work by our group on TS-1, including the tuning of coordination states of Ti, improvement of diffusion properties, and industrial applications of the HPPO route. We hope to provide some references for related research.

Acknowledgements

The authors acknowledge financial support from the National Key Research and Development Program of China (2016YFB0301704), the National Natural Science Foundation of China (21506021), and the Fundamental Research Funds for the Central Universities (DUT19LK61).

Conflict of interest

There is no conflict of interest to be declared.

Author details

Yi Zuo, Min Liu and Xinwen Guo*
State Key Laboratory of Fine Chemicals, PSU-DUT Joint Center for Energy Research, Department of Catalysis Chemistry and Engineering, School of Chemical Engineering, Dalian University of Technology, Dalian, China

*Address all correspondence to: guoxw@dlut.edu.cn

References

[1] Rouquerol J, Avnir D, Fairbridge CW, Everett DH, Haynes JM, Pernicone N, et al. Recommendations for the characterization of porous solids. Pure and Applied Chemistry. 1994;**66**:1739-1758. DOI: 10.1351/pac199466081739

[2] Chen S, Wang Q. Scientific Fundamentals and Technologies for Preparation of Solid Catalysts. Beijing: Chemical Industry Press; 2012. pp. 110-111. DOI: 9787122147165

[3] Taramasso M, Perego G, Notari B. Preparation of porous crystalline synthetic materials comprised of silicon and titanium oxides. US Patent. 1983;4410501

[4] Clerici MG, Bellussi G, Romano U. Synthesis of propylene oxide from propylene and hydrogen peroxide catalyzed by titanium silicalite. Journal of Catalysis. 1991;**129**:159-167. DOI: 10.1016/0021-9517(91)90019-Z

[5] Zhang Z, Kong XX, Feng M, Luo ZH, Lu H, Cao GP. In situ synthesis of TS-1 on carbon nanotube decorated nickel foam with ultrafine nanoparticles and high content of skeleton titanium. Industrial and Engineering Chemistry Research. 2019;**58**:69-78. DOI: 10.1021/acs.iecr.8b04545

[6] Xiong G, Hu D, Guo Z, Meng Q, Liu L. An efficient titanium silicalite-1 catalyst for propylene epoxidation synthesized by a combination of aerosol-assisted hydrothermal synthesis and recrystallization. Microporous and Mesoporous Materials. 2018;**268**:93-99. DOI: 10.1016/j.micromeso.2018.04.015

[7] Li Y, Fan Q, Li Y, Feng X, Chai Y, Liu C. Seed-assisted synthesis of hierarchical nanosized TS-1 in a low-cost system for propylene epoxidation with H_2O_2. Applied Surface Science. 2019;**483**:652-660. DOI: 10.1016/j.apsusc.2019.03.334

[8] Wang B, Han H, Ge B, Ma J, Zhu J, Chen S. An efficient hydrophobic modification of TS-1 and its application in the epoxidation of propylene. New Journal of Chemistry. 2019;**43**:10390-10397. DOI: 10.1039/c9nj01937e

[9] Liu X, Liu J, Xia Y, Yin D, Steven RK, Mao L. Catalytic performance of TS-1 in oxidative cleavage of 1-alkenes with H_2O_2. Catalysis Communications. 2019;**126**:40-43. DOI: 10.1016/j.catcom.2019.04.021

[10] Martens JA, Buskens P, Jacobs PA, van der Pol A, van Hooff JHC, Ferrini C, et al. Hydroxylation of phenol with hydrogen peroxide on EUROTS-1 catalyst. Applied Catalysis A: General. 1993;**99**:71-84. DOI: 10.1016/0926-860X(93)85040-V

[11] Wu W, Tran DT, Wu X, Oh SC, Wang M, Chen H, et al. Multilamellar and pillared titanium silicalite-1 with long-range order of zeolite nanosheet layers: Synthesis and catalysis. Microporous and Mesoporous Materials. 2019;**278**:414-422. DOI: 10.1016/j.micromeso.2019.01.010

[12] Shen X, Wang J, Liu M, Li M, Lu J. Preparation of the hierarchical Ti-rich TS-1 via tritonX-100-assisted synthetic strategy for the direct oxidation of benzene. Catalysis Letters. 2019;**149**:2586-2596. DOI: 10.1007/s10562-019-02735-5

[13] Mantegazza MA, Leofanti G, Petrini G, Padovan M, Zecchina A, Bordiga S. Selective oxidation of ammonia to hydroxylamine with hydrogen peroxide on titanium based catalysts. Studies in Surface Science and Catalysis. 1994;**82**:541-550. DOI: 10.1016/S0167-2991(08)63447-3

[14] Hu Y, Dong C, Wang T, Luo G. Cyclohexanone ammoximation over TS-1 catalyst without organic solvent in a microreaction system. Chemical Engineering Science. 2018;**187**:60-66. DOI: 10.1016/j.ces.2018.04.044

[15] Xue Y, Zuo G, Wen Y, Wei H, Liu M, Wang X, et al. Seed-assisted synthesis of TS-1 crystals containing Al with high catalytic performances in cyclohexanone ammoximation. RSC Advances. 2019;**9**:2386-2394. DOI: 10.1039/c8ra10104c

[16] Huybrechts DRC, De Bruycker L, Jacobs PA. Oxyfunctionalization of alkanes with hydrogen peroxide on titanium silicalite. Nature. 1990;**345**:240-242. DOI: 10.1038/345240a0

[17] Zhu M, Zhu C, Wu D, Wang X, Wang H, Gao J, et al. Efficient photocatalytic water splitting through titanium silicalite stabilized CoO nanodots. Nanoscale. 2019;**11**:15984-15990. DOI: 10.1039/c9nr05057d

[18] Du Q, Guo Y, Wu P, Liu H, Chen Y. Facile synthesis of hierarchical TS-1 zeolite without using mesopore templates and its application in deep oxidative desulfurization. Microporous and Mesoporous Materials. 2019;**275**:61-68. DOI: 10.1016/j.micromeso.2018.08.018

[19] Chen R, Liu C, Johnson NW, Zhang L, Mahendra S, Liu Y, et al. Removal of 1,4-dioxane by titanium silicalite-1: Separation mechanisms and bioregeneration of sorption sites. Chemical Engineering Journal. 2019;**371**:193-202. DOI: 10.1016/j.cej.2019.03.285

[20] Millini R, Massara EP, Perego G, Bellussi G. Framework composition of titanium silicalite-1. Journal of Catalysis. 1992;**137**:497-503. DOI: 10.1016/0021-9517(92)90176-I

[21] Carati A, Flego C, Previde Massara E, Millini R, Carluccio L, Parker WO Jr, et al. Stability of Ti in MFI and Beta structures: A comparative study. Microporous and Mesoporous Materials. 1999;**30**:137-144. DOI: 10.1016/S1387-1811(99)00018-9

[22] Clerici MG, Ingallina P. Epoxidation of lower olefins with hydrogen peroxide and titanium silicalite. Journal of Catalysis. 1993;**140**:71-83. DOI: 10.1006/jcat.1993.1069

[23] Yoon CW, Hirsekorn KF, Neidig ML, Yang XZ, Tilley TD. Mechanism of the decomposition of aqueous hydrogen peroxide over heterogeneous TiSBA15 and TS-1 selective oxidation catalysts: Insights from spectroscopic and density functional theory studies. ACS Catalysis. 2011;**1**:1665-1678. DOI: 10.1021/cs2003774

[24] Guo Q, Sun K, Feng Z, Li G, Guo M, Fan F, et al. A thorough investigation of the active titanium species in TS-1 zeolite by in situ UV resonance Raman spectroscopy. Chemistry--A European Journal. 2012;**18**:13854-13860. DOI: 10.1002/chem.201201319

[25] Wang L, Xiong G, Su J, Li P, Guo H. In situ UV Raman spectroscopic study on the reaction intermediates for propylene epoxidation on TS-1. Journal of Physical Chemistry C. 2012;**116**:9122-9131. DOI: 10.1021/jp3017425

[26] Wang L, Wang X, Guo X, Li G, Xiu J. Quick synthesis of titanium silicalite-1. Chinese Journal of Catalysis. 2001;**22**:513-514. DOI: 0253-9837(2001)06-0513-02

[27] Zuo Y, Wang X, Guo X. Synthesis of titanium silicalite-1 with small crystal size by using mother liquid of titanium silicalite-1 as seed. Industrial and Engineering Chemistry Research. 2011;**50**:8485-8491. DOI: 10.1021/ie200281v

[28] Zuo Y, Liu M, Zhang T, Hong L, Guo X, Song C, et al. Role of pentahedrally coordinated titanium in titanium silicalite-1 in propene epoxidation. RSC Advances. 2015;**5**:17897-17904. DOI: 10.1039/c5ra00194c

[29] Fan WB, Duan RG, Yokoi T, Wu P, Kubota Y, Tatsumi T. Synthesis, crystallization mechanism, and catalytic properties of titanium-rich TS-1 free of extraframework titanium species. Journal of the American Chemical Society. 2008;**130**:10150-10164. DOI: 10.1021/ja7100399

[30] Wang J, Zhao YL, Yokoi T, Kondo JN, Tatsumi T. High-performance titanosilicate catalyst obtained through combination of liquid-phase and solid-phase transformation mechanisms. ChemCatChem. 2014;**6**:2719-2726. DOI: 10.1002/cctc.201402239

[31] Wang YM, He JQ. The method of adjusting pH value by oligosaccharide in the synthesis process of titanium silicalite. CN Patent. 2013;201210048648.7

[32] Zhang T, Zuo Y, Liu M, Song C, Guo X. Synthesis of titanium silicalite-1 with high catalytic performance for 1-butene epoxidation by eliminating the extraframework Ti. ACS Omega. 2016;**1**:1034-1040. DOI: 10.1021/acsomega.6b00266

[33] Zuo Y, Song W, Dai C, He Y, Wang M, Wang X, et al. Modification of small-crystal titanium silicalite-1 with organic bases: Recrystallization and catalytic properties in the hydroxylation of phenol. Applied Catalysis A: General. 2013;**453**:272-279. DOI: 10.1016/j.apcata.2012.12.027

[34] Zuo Y, Wang X, Guo X. Synthesis of titanium silicalite-1 with small crystal size by using mother liquor of titanium silicalite-1 as seeds (II): Influence of synthesis conditions on properties of titanium silicalite-1. Microporous and Mesoporous Materials. 2012;**162**:105-114. DOI: 10.1016/j.micromeso.2012.06.016

[35] Zuo Y, Liu M, Zhang T, Meng C, Guo X, Song C. Enhanced catalytic performance of titanium silicalite-1 in tuning the crystal size in the range 1200-200 nm in a tetrapropylammonium bromide system. ChemCatChem. 2015;**7**:2660-2668. DOI: 10.1002/cctc.201500440

[36] Wang Y, Zuo Y, Liu M, Dai C, Feng Z, Guo X. The high-performance hollow silicalite-1@titanium silicalite-1 core-shell catalyst for propene epoxidation. ChemistrySelect. 2017;**2**:10097-10100. DOI: 10.1002/slct.201701753

[37] Zuo Y, Liu M, Hong L, Wu M, Zhang T, Ma M, et al. Role of supports in the tetrapropylammonium hydroxide treated titanium silicalite-1 extrudates. Industrial and Engineering Chemistry Research. 2015;**54**:1513-1519. DOI: 10.1021/ie504531v

[38] Zuo Y, Liu M, Ma M, Wang Y, Guo X, Song C. Enhanced catalytic activity on post-synthesized hollow titanium silicalite-1 with high titanium content on the external surface. ChemistrySelect. 2016;**1**:6160-6166. DOI: 10.1002/slct.201601430

[39] Jacobsen CJH, Madsen C, Houzvicka J, Schmidt I, Carlsson A. Mesoporous zeolite single crystals. Journal of the American Chemical Society. 2000;**122**:7116-7117. DOI: 10.1021/ja000744c

[40] Zuo Y, Zhang T, Liu M, Ji Y, Song C, Guo X. Mesoporous/microporous titanium silicalite with controllable pore diameter for cyclohexene epoxidation. Industrial and Engineering Chemistry Research. 2018;**57**:512-520. DOI: 10.1021/acs.iecr.7b03719

[41] Zuo Y, Wang M, Song W, Wang X, Guo X. Characterization and catalytic performance of deactivated and regenerated TS-1 extrudates in a pilot plant of propene epoxidation. Industrial and Engineering Chemistry Research. 2012;**51**:10586-10594. DOI: 10.1021/ie300581z